AF662750

Classical Methods
Volume 1

Wiley India Pvt. Ltd.

Analytical Chemistry by Open Learning

Titles in Series:

1. Samples and Standards
2. Sample Pretreatment and Separation
3. Classical Methods Vol. 1
4. Classical Methods Vol. 2
5. Measurement, Statistics and Computation
6. Using Literature
7. Instrumentation
8. Chromatographic Separations
9. Gas Chromatography, 2nd Edition
10. High Performance Liquid Chromatography, 2nd Edition
11. Electrophoresis
12. Thin Layer Chromatography
13. Ultraviolet and Visible Spectroscopy
14. Fluorescence and Phosphorescence Spectroscopy
15. Biological Applications of Infra Red Spectroscopy
16. Atomic Absorption and Plasma Spectroscopy, 2nd Edition
17. Nuclear Magnetic Resonance Spectroscopy
18. X-Ray Methods
19. Mass Spectrometry, 2nd Edition
20. Scanning Electron Microscopy and X-Ray Microanalysis
21. Principles of Electroanalytical Methods
22. Potentiometry and Ion Selective Electrodes
23. Polarography and Other Voltammetric Methods
24. Radiochemical Methods
25. Clinical Specimens
26. Diagnostic Enzymology
27. Quantitative Bioassay
28. Assessment and Control of Biochemical Methods
29. Thermal Methods
30. Microprocessor Applications
31. Environmental Analysis
32. Quality in the Analytical Chemistry Laboratory

Classical Methods
Volume 1

Analytical Chemistry by Open Learning

Authors:
DEREK COOPER
North Staffordshire Polytechnic

CHRIS DORAN
Wirral Metropolitan College

Editor:
NORMAN B. CHAPMAN

on behalf of ACOL

Published on behalf of ACOL, London
by
JOHN WILEY & SONS
Chichester • New York • Brisbane • Toronto • Singapore

Classical Methods, Volume1

Published by permission of the Controller of Her Majesty's Stationery Office

Authorized reprint by Wiley India Pvt. Ltd., 4435/7, Ansari Road, Daryaganj, New Delhi – 110002.

Reprint: 2008

Printed at: *Ar Emm International, New Delhi*

ISBN: 978-81-265-1713-8

Analytical Chemistry

This series of texts is a result of an initiative by the Committee of Heads of Polytechnic Chemistry Departments in the United Kingdom. A project team based at Thames Polytechnic using funds available from the Manpower Services Commission 'Open Tech' Project have organised and managed the development of the material suitable for use by 'Distance Learners'. The contents of the various units have been identified, planned and written almost exclusively by groups of polytechnic staff, who are both expert in the subject area and are currently teaching in analytical chemistry.

The texts are for those interested in the basics of analytical chemistry and instrumental techniques who wish to study in a more flexible way than traditional institute attendance or to augment such attendance. A series of these units may be used by those undertaking courses leading to BTEC (levels IV and V), Royal Society of Chemistry (Certificates of Applied Chemistry) or other qualifications. The level is thus that of Senior Technician.

It is emphasised however that whilst the theoretical aspects of analytical chemistry can be studied in this way there is no substitute for the laboratory to learn the associated practical skills. In the U.K. there are nominated Polytechnics, Colleges and other Institutions who offer tutorial and practical support to achieve the practical objectives identified within each text. It is expected that many institutions worldwide will also provide such support.

The project will continue at Thames Polytechnic to support these 'Open Learning Texts', to continually refresh and update the material and to extend its coverage.

Further information about nominated support centres, the material or open learning techniques may be obtained from the project office at Thames Polytechnic, ACOL, Wellington St., Woolwich, London, SE18 6PF.

How to Use an Open Learning Text

Open learning texts are designed as a convenient and flexible way of studying for people who, for a variety of reasons cannot use conventional education courses. You will learn from this text the principles of one subject in Analytical Chemistry, but only by putting this knowledge into practice, under professional supervision, will you gain a full understanding of the analytical techniques described.

To achieve the full benefit from an open learning text you need to plan your place and time of study.

- Find the most suitable place to study where you can work without disturbance.
- If you have a tutor supervising your study discuss with him, or her, the date by which you should have completed this text.
- Some people study perfectly well in irregular bursts, however most students find that setting aside a certain number of hours each day is the most satisfactory method. It is for you to decide which pattern of study suits you best.
- If you decide to study for several hours at once, take short breaks of five or ten minutes every half hour or so. You will find that this method maintains a higher overall level of concentration.

Before you begin a detailed reading of the text, familiarise yourself with the general layout of the material. Have a look at the course contents list at the front of the book and flip through the pages to get a general impression of the way the subject is dealt with. You will find that there is space on the pages to make comments alongside the text as you study—your own notes for highlighting points that you feel are particularly important. Indicate in the margin the points you would like to discuss further with a tutor or fellow student. When you come to revise, these personal study notes will be very useful.

Π When you find a paragraph in the text marked with a symbol such as is shown here, this is where you get involved. At this point you are directed to do things: draw graphs, answer questions, perform calculations, etc. Do make an attempt at these activities. If necessary cover the succeeding response with a piece of paper until you are ready to read on. This is an opportunity for you to learn by participating in the subject and although the text continues by discussing your response, there is no better way to learn than by working things out for yourself.

We have introduced self assessment questions (SAQ) at appropriate places in the text. These SAQs provide for you a way of finding out if you understand what you have just been studying. There is space on the page for your answer and for any comments you want to add after reading the author's response. You will find the author's response to each SAQ at the end of the text. Compare what you have written with the response provided and read the discussion and advice.

At intervals in the text you will find a List of Objectives. This will emphasise the important points covered by the material you have just read and will give you a checklist of tasks you should then be able to achieve.

You can revise the Unit, perhaps for a formal examination, by re-reading the Objectives, and by working through some of the SAQs. This should quickly alert you to areas of the text that need further study.

At the end of the book you will find for reference lists of commonly used scientific symbols and values, units of measurement and also a periodic table.

Contents

Study Guide

Chemical analysis has been of interest to mankind for many centuries, although in its infancy it was not always recognised as such. Records dating from the early civilisations of Egypt, Greece, and Rome indicate a very pronounced interest in the purity of precious metals, and in ways of detecting adulteration. At a more everyday level people have long been interested in the strength and quality of beers and wines, and of course, from the later middle ages there was a very pressing requirement to maintain the purity of the components of gunpowder. The massive expansion of trade and commerce which accompanied the industrial revolution brought along with it a multiplicity of new materials for which analytical information was needed.

During the late nineteenth century there was a concentrated effort to formalise the work of analysts, and indeed to secure the role of the analyst as a professional one. The methods which were developed during this era are all the so called 'wet methods' or 'classical methods' of analysis. These methods are often quick, cheap and reliable and still find wide application in industry. In modern parlance they are distinguished from 'instrumental methods' of analysis of which the meaning is self evident. This unit deals with modern versions of the 'classical methods'.

Classical methods of analysis may be applied to virtually every type of analyte encountered in industry and commerce. However, there are groups of compounds for which classical methods are not particularly appropriate. The classical methods are typically used for monitoring purity (at the high percentage level) and checking impurities (at the low percentage level). They are less useful for chemical trace analysis and for pharmaceuticals, organic fine chemicals, petroleum, and plastics.

The Unit contains thirteen Parts and is presented in two volumes. These Parts fall broadly into three categories. The first category is a treatment of the equilibrium model and its application to classical analysis. This covers Parts 1 to 5. The second category deals with volumetric analysis, that is analysis in which the principal measured parameters are volumes. The use of pipettes and burettes in titrations is the most characteristic activity here. Parts 6, 7, 10, 11, 12 and 13 cover these aspects. The major subdivisions are neutralisation titrimetry (7); redox titrimetry (11); and complexometric titrimetry (12 and 13). A small section on precipitation titrations, 10, is also included. The remaining category utilises measurement by weighing as the principal activity, and the technique is known as gravimetric analysis. Parts 8 and 9 cover this topic.

Part 1 develops the idea of chemical equilibrium as used by analytical chemists and the basic concept or model is then applied to major features of practical interest to the analyst: ionisation Part 2, solubility Part 3, complex formation Part 4, and redox systems Part 5.

You should aim to work through Part 1 to 5 before studying the other Parts. This will give you a secure foundation for understanding the variety of practical points which are raised in later Parts. After completing Parts 1 to 5 you will also be confident in your approach to the range of calculations associated with analytical results.

Part 6 is essential to all further study of volumetric methods, but following Part 6 you may move straight on to Part 7 or follow Part 6 by Part 10, or 11, or 12 plus 13. Whichever route you decide to use you will find it useful to re-read the appropriate equilibrium Part before the more practically biased exposition: ie Part 2 precedes 7; Part 3 precedes 9; Part 4 precedes 12–13 and Part 5 precedes 11.

Preknowledge

Part 1 of this Unit contains introductory material, some of it may be familiar to you from earlier studies. The particular emphasis is on developing a model for equilibria, and on using concepts of equilibrium systems as they relate to chemical analysis.

The preknowledge assumed is as follows.

- Competence in writing balanced chemical equations to represent chemical processes.

- The ability to make calculations based on stoichiometric relationships.

- Familiarity with the concept of *molar concentration.*

- Familiarity with a simple model for the ionisation of electrolytes, and solvation in solution.

- Acquaintanceship with the concepts of pH, redox processes, and the coordination model of complex-ion formation.

Practical Objectives

Parts 1–6 of Classical Methods deal with the development and the application of the equilibrium model to all branches of classical analysis. Because of this the practical activity associated with the theoretical discussion in Parts 2 to 5 are subsumed in the appropriate part dealing with practical applications, ie Parts 7 to 13.

The expected learning outcome on completion of this Unit is that the student can:

— use the analytical balance with competence and understand the circumstances in which precise weighing is necessary;

— calibrate volumetric glassware and understand cleaning procedures;

— prepare standard solutions from primary standards and standardise other solutions;

— carry out replicate neutralisation titrations and use several indicators;

— carry out pH titrations (with or without computer assisted processing);

— determine the purity of several monobasic and polybasic acids and the composition of mixtures thereof;

- prepare buffer solutions and apply them in an analysis;
- carry out a neutralisation titration in a non-aqueous medium;
- select correct conditions for precipitation;
- select correct filtration media and drying methods;
- use adsorption indicators in a precipitation titration;
- carry out the gravimetric determination of two model analytes;
- use the platinum indicator electrode to perform a redox titration;
- carry out representative analyses by using permanganate, cerate and dichromate solutions;
- use a selection of other oxidants for the determination of analytes by redox titrations;
- determine a single metal-ion by a complexometric method;
- analyse mixtures of metal ions by using EDTA;
- use masking agents in compleximetry;
- carry out several 'real-world' classical analyses chosen from an application field of direct interest to the student.

Bibliography

There are many text books available which cover the subject of Classical Analysis. The following are just a small sample of useful texts dealing with the introductory concepts in Part 1.

1. D McQuarrie and P Rock, *General Chemistry*, Chapters 15, 17, 18, 19, 20, Freeman, 1984.

2. W Masterton, E J Slowinski and C L Stanitski, *Chemical Principles*, 6th Edition, Chapters 19, 20, 21, 23, 24, Saunders, 1985.

Analytical texts covering the material of Parts 2, 3, 4 and 5.

3. R W Ramette, *Chemical Equilibrium and Analysis*, Addison Wesley, 1981.

4. D A Skoog and D M West, *Fundamentals of Analytical Chemistry*, Chapters 2, 3, 15, Holt, Rinehart and Winston, 1976.

The following books contain useful chapters on Volumetric Analysis, dealt with in Part 6 and 7.

5. L F Hamilton and S G Simpson, *Quantitative Chemical Analysis*, Macmillan, 1964.

6. C T Kenner and K W Busch, *Quantitative Analysis*, Macmillan, 1979.

7. R A Day and A L Underwood, *Quantitative Analysis*, Prentice Hall, 1967.

8. R D Braun, *Introduction to Chemical Analysis*, McGraw Hill, 1982.

9. G D Christian, *Analytical Chemistry*, Wiley, 1980.

Good standard works for future study.

10. I M Kolthoff and P J Elving, *Treatise on Analytical Chemistry*, Pt 1, Vol I, Section B, Chapters 7–19, Wiley, 1975.

11. I M Kolthoff and P J Elving, *Treatise on Analytical Chemistry*, Pt I, Vol II, Section I, Wiley, 1975.

Acknowledgements

Figures 7.5d and 7.5e are redrawn from G. D. Christian, *Analytical Chemistry*, John Wiley & Sons, New York, 1980 with permission.

1. Equilibria and Equilibrium Systems

This Part describes equilibria in simple qualitative terms and relates the concept to everyday experience. The dynamic nature of equilibria is illustrated and the response to external changes is described. This is followed by a description of the limitations imposed on the model in its application in analytical practice. With these limitations in mind the theoretical basis of the model is described. This is followed by illustrations of features of interest in analytical practice. The Part is concluded with a brief survey of the phenomena to which we shall apply the model in subsequent Parts.

1.1. REVERSIBILITY AND DYNAMIC INTERACTION

1.1.1. Equilibria and Equilibrium Systems

We have all met the term 'equilibrium' as used in common parlance. It is associated with 'balance' and 'stability'. The word itself is derived from the prefix *equi* meaning equal, and the Latin word *libra* for scales or balance. We must be careful to separate this general meaning of 'equilibrium' from its scientific meaning. An analogy will be used to illustrate this before we give a more detailed account and attempt to make the ideas quantitative.

We take the analogy of a tight-rope walker with a balance pole fitted with weights at both ends. If the masses at each end are equal the system is in equilibrium. If we now add weights to one side, the system becomes unbalanced and equilibrium is lost. The tight-rope walker must restore equilibrium by leaning over; otherwise he falls beyond the point at which recovery is possible. The chemical equilibria with which we are concerned in this section are similar to the behaviour of the tight-rope walker in one sense. Just as changes in the tight-rope walker's distribution of weights requires a change in his posture to maintain his equilibrium, so *perturbation of a chemical equilibrium system leads to changes in the distribution of the components*. However unlike the tight-rope walker, who may fall off his wire, the changes in the chemical system always occur in the manner necessary to *restore* equilibrium.

The reason this is possible is that the systems which are properly described as chemical equilibria are *reversible*, ie they are dynamically interactive. This reversibility is conventionally represented by a double arrow; thus chemical equilibria are typically represented as follows.

$$A + B \rightleftarrows C + D$$

As written, we call A and B the *reactants* and C and D the *products*. But of course the system is reversible, consequently we might just as well represent it as:

$$C + D \rightleftarrows A + B$$

An important thing to recognise at this stage is that these equations do *not* tell us anything about 'how far the reaction has gone', ie we know nothing as yet about the distribution of the component material between A, B, C and D.

SAQ 1.1a

Which of the following represent the same equilibrium?

(*i*) $NH_{3(aq)} + H_2O_{(l)} \rightleftharpoons NH^+_{4(aq)} + OH^-_{(aq)}$

(*ii*) $NH_{3(aq)} + H_3O^+_{(aq)} \rightleftharpoons NH^+_{4(aq)} + H_2O_{(l)}$

(*iii*) $NH^+_{4(aq)} + OH^-_{(aq)} \rightleftharpoons NH_{3(aq)} + H_2O_{(l)}$

(*iv*) $H_2PO^-_{4(aq)} + H_2O_{(l)} \rightleftharpoons H_3O^+_{(aq)} + HPO^{2-}_{4(aq)}$

(*v*) $HPO^{2-}_{4(aq)} + H_3O^+_{(aq)} \rightleftharpoons H_2PO^-_{4(aq)} + H_2O_{(l)}$

(*vi*) $H_2PO^-_{4(aq)} + OH^-_{(aq)} \rightleftharpoons HPO^{2-}_{4(aq)} + H_2O_{(l)}$

Notice that in the SAQ, subscripts have been used to indicate that some of the species are dissolved in water or *aquated*, (aq), and that the water is present as a *liquid*, (l). The subscript (s) is similarly used to represent the presence of a *solid*. In many systems with which the analyst is concerned it is generally understood that we are dealing with species in solution, and that water is in the liquid phase. Consequently these subscripts are frequently omitted. It is however necessary to have regard for this convention when dealing with, for example, solids in equilibrium with saturated solutions.

$$\text{eg} \quad AgCl_{(s)} \rightleftharpoons Ag^+_{(aq)} + Cl^-_{(aq)}$$

The double arrow simply means that the process is at equilibrium. It does not imply that the system is predominantly made up of the species on one side or the other. However, a convention which you may occasionally meet, uses a combination of a solid arrow and a dotted arrow, or sometimes a thin arrow and a thick arrow. This is intended to indicate in a general way that the equilibrium lies strongly to one side.

$$\text{eg} \quad A + B \rightleftharpoons C + D$$

$$\text{or} \quad A + B \rightleftharpoons C + D$$

These mean that the equilibrium mixture is principally C and D with only traces of A and B. You will see later in this section why the use of both arrows solid, and a more quantitative expression is preferable.

We shall pursue this qualitative model a little further, to identify features which help us to understand the part played by equilibria in analytical chemistry. It is worth noting that in a strict theoretical sense, equilibrium systems should be *closed systems*, ie matter should not be allowed to enter or leave the system. In practice the losses from systems in solution chemistry are so small that they can be ignored for most analytical applications.

1.1.2. The System is Dynamic

We now take a general equilibrium system represented by A and B as reactants and C and D as products. We start with 100% of A and B. Reaction occurs so that the proportions of A and B decrease, while those of C and D increase. Eventually the *proportions* no longer change and the system is said to be in equilibrium. The question is – 'Has the reaction stopped simply because the quantities of each component are no longer changing?'. If the system is truly reversible we might expect some C and D to be reacting to form A and B, while at the same time the forward reaction is occurring.

This dynamic reversibility may be illustrated by examining the behaviour of normal and radio-labelled potassium chloride in solution. When potassium chloride is added to water at 273 K (0 °C) the solid dissolves until about 29 g per 100 g of water have dissolved. Further addition of potassium chloride appears to have no effect on the mixture and the solution is said to be saturated. However, if we now add solid KCl containing radio-labelled potassium, there is no visible change, but monitoring of the solution shows a slow transfer of radioactivity into the solution. Analytical measurements of total potassium in solution show that this is constant, thus our observation means that some potassium ions must be leaving the solution and becoming part of the solid at the same time as the radio-labelled potassium is entering the solution. This dynamic reversibility is general for equilibrium systems.

SAQ 1.1b

The expression below represents a general equilibrium system.

$$A + B \rightleftharpoons C + D$$

Mark the following statements as either true or false. ⟶

SAQ 1.1b (cont.)

'The use of double arrows indicates that ...

(*i*) ... a mixture of reactants A and B reacts almost completely to give largely products, C and D.'

true ... false ...

(*ii*) ... a mixture of reactants C and D reacts almost completely to give largely products A and B.'

true ... false ...

(*iii*) ... if the reactants are A and B, then reaction takes place completely to give C and D, but if the starting materials are C and D, then complete reaction takes place to give A and B.'

true ... false ...

(*iv*) ... the reaction takes place smoothly and rapidly.'

true ... false ...

(*v*) ... the reaction is reversible.'

true ... false ...

(*vi*) ... there are approximately equal amounts of A and B and C and D present when the reaction reaches equilibrium.'

true ... false ...

(*vii*) ... the system contains both A and B (some of which is undergoing reaction to C and D), and C and D (some of which is undergoing reaction to A and B).'

true ... false ...

⟶

SAQ 1.1b (cont.)

(*viii*) ... the system can just as well be written as C + D $\rightleftarrows$ A + B.'

true ... false ...

(*ix*) ... the reaction is not stoichiometric.'

true ... false ...

(*x*) ... it is not possible to determine by experiment whether the starting materials were A + B or C + D.'

true ... false ...

1.1.3. Perturbation of the Components

Again we shall take our general system as given below.

$$A + B \rightleftarrows C + D$$

Recall the analogy of the tight-rope walker: what happens if we remove part of one component, say D? The system responds by allowing more A and B to react to produce more of D. A new equilibrium is established with different proportions of the components. If we now replace the component D the system responds by forming more A and B again. This feature will be developed quantitatively later in the Part.

1.1.4. The Effect of Temperature

Returning to our general equilibrium system of A and B in equilibrium with C and D, it is found experimentally that the *relative* quantities of A and B, and C and D are constant *at a constant temperature*. If the temperature is changed then the relative quantities change until a new equilibrium is reached. That is, the *position* of the equilibrium depends on temperature.

1.1.5. Limitations

In principle it should be possible to describe all chemical reactions as equilibria as long as one insists that the system is *closed*. A closed system in a material sense is one which does not allow matter either to enter or to leave the system. However interesting this may be from theoretical view-point, it is not particularly helpful for 'real-world' applications of equilibrium concepts.

It is obvious that in practice many systems of interest are not closed. Even many highly precise analytical manipulations are carried out with aqueous solutions in open containers, thus permitting loss of solvent water. However these losses are sufficiently small for the system to be regarded as effectively closed.

Another apparent conflict with theory occurs when we consider reactions which are so strongly exothermic that they have essentially 'gone to completion'. Thus we must, in practice, qualify our range of processes to which equilibrium concepts are applicable by restricting them to processes which are *detectably reversible*.

A further limitation to the concept is that it has nothing to say about the speed with which changes occur. A change may be thermodynamically reversible, ie infinitesimal reversal of the constraints reverses the direction of the change, but the change itself may be either rapid or slow. The analyst usually exploits changes which are rapid. Indeed he may take steps to ensure that reactions are as fast as may be convenient, if there is any chance of their being unacceptably slow. There are however many processes which are so slow that true thermodynamic equilibrium is not attained. These processes may *appear* to be at equilibrium but in fact they may simply be changing at an undetectably low rate. They are not adequately described by the equilibrium model which we shall develop.

SAQ 1.1c

Several chemical systems are described below.

For each of these state whether the system is usefully described as an equilibrium system. For those processes which you identify as equilibria, write one or more equations which describe the system.

(*i*) Solid iodine in contact with iodine vapour in a closed container.

(*ii*) A sample of radio-labelled potassium iodide undergoing radioactive decay.

(*iii*) A solution of ethanoic acid (acetic acid) in water.

(*iv*) A sample of oxalic (ethanedioic) acid solution in an open flask being oxidised at 363 K (90 °C) according to the equation below.

$$2\,MnO_4^- + 5\,H_2C_2O_4 + 6\,H^+ \rightleftharpoons 2\,Mn^{2+} + 10\,CO_2 + 8\,H_2O$$

(*v*) A solution of ethanoic acid in 'heavy' water, D_2O.

(*vi*) A small volume of a dilute solution of iodine in tetrachloromethane (carbon tetrachloride) in contact with and enclosed by a larger volume of water, in an open flask.

(*vii*) A mixture of aqueous copper(II) sulphate and concentrated hydrochloric acid.

SAQ 1.1c

1.1.6. A Practical Problem

One of the problems in dealing with equilibrium systems is the relationship between *concentration* and *activity*. In the laboratory we are principally interested in concentrations; this is one of the parameters which are readily accessible practically. However the strict theoretical treatment of equilibria requires the use of activities. You may have met these terms in previous study but a general analogy to illustrate the difference is useful here.

We have a large wood-working shop. One carpenter in the workshop can make tables at a particular rate. Two carpenters in the workshop will produce tables at twice the rate of the single man, similarly three carpenters at three times the rate. However the workshop eventually becomes moderately crowded and the men begin to obstruct one another's work; additional carpenters do not produce the same increase in output at this stage as they did originally. Thus at high levels of occupancy, production is not proportional to the number of carpenters. Another way of saying this is that the *activity* of the carpenters is proportional to the *concentration* of carpenters when their concentration is low, but activity increasingly deviates from this at higher values of concentration.

The same is true of species in solution. The activity is virtually the same as concentration at low concentrations, ie at high dilution, but this assertion becomes less valid in more concentrated solutions.

We normally relate activity, a, and concentration, c by the activity coefficient, f.

$$a = fc$$

Activity is the term we must use for precise applications of the equilibrium concept, whereas concentration is the parameter over which we have most control. For dilute solutions f approaches unity, so we can equate activity and the molar concentration. Molar concentrations are indicated by the symbol for, or formula of the material enclosed in square brackets. Thus $[Cl^-][Ag^+]$ means the numerical value of the molar concentration of chloride ions multiplied by the numerical value of the molar concentration of silver ions.

1.2. REPRESENTATION

1.2.1. Theoretical Basis

We return to our system in which A and B are in equilibrium with C and D, the relative concentrations of the components are unchanging. We should emphasise the word *relative* here. The distribution of material between the species A, B, C and D remains the same even if all concentrations are multiplied by a factor of 10. However we know that the A's and B's are still reacting to give C's and D's, and *vice-versa*, as the system is dynamic and reversible. Therefore it follows that at equilibrium the *rate* of the reaction between A and B must be equal to the *rate* of the reaction between C and D. This was clearly not true in the early stages of the reaction when we had largely A and B. Thus it follows that the rate of the reaction must be related to the amount of material present. These ideas were given quantitative expression over a century ago by Guldberg and Waage in the *Law of Mass Action*.

1.2.2. The Law of Mass Action

The rate of a reaction, at a fixed temperature, is proportional to the product of the activities (active concentrations) of the reacting substances, each raised to the power numerically equal to the number of molecules of each appearing in the balanced stoichiometric equation. We have already noted that it is generally quite acceptable to use molar concentrations in place of activities for low concentrations.

For our basic equilibrium,

$$A + B \rightleftarrows C + D$$

the Law of Mass Action relates the forward reaction rate v_1, to the concentrations thus.

$$v_1 \propto [A][B]$$

Conventionally we convert this into an equality by means of a constant, here called the rate constant, k_1.

$$v_1 = k_1[A][B]$$

A similar treatment for the reverse reaction gives the following expression.

$$v_2 = k_2[C][D]$$

When the system is at equilibrium, it is clear that the rate of the forward reaction must be equal to that of the reverse reaction,

ie $v_1 = v_2$ (at the temperature of the experiment).

Thus $$k_1[A][B] = k_2[C][D]$$

or $$\frac{k_1}{k_2} = \frac{[C][D]}{[A][B]}$$

In practice the determination of rate constants is not at all straightforward; however the equilibrium concentrations themselves, and consequently the ratio k_1/k_2 are usually readily determined. We are principally concerned with the value of this ratio and will refer to it as the *Equilibrium Constant*, K_{eq}.

$$K_{eq} = \frac{k_1}{k_2} = \frac{[C][D]}{[A][B]}$$

The convention for K_{eq} is that the right-hand side of the chemical equation enters into the numerator and the left-hand side the denominator.

SAQ 1.2a

How is the value of K_{eq} for the equilibrium (*i*) $C + D \rightleftarrows A + B$ related to that for the equilibrium (*ii*) $A + B \rightleftarrows C + D$?

Note that because K_{eq} is arrived at *via* a ratio of rate constants, high values of K_{eq} do *not* necessarily imply high values of the rate constants themselves. It is not possible to obtain any information about rates simply by measuring K_{eq}. You will also notice that this relationship between the ratio of rate constants and K_{eq} indicates a temperature dependence of K_{eq} values. This is because the rate constants change to different degrees with changing temperature.

SAQ 1.2b

Indicate which of the following are correct completions of the statement:

'The law of mass action states that ...

(*i*) ... the rates at which equilibrium processes proceed are proportional to the indices which appear in the balanced equation.'

(*ii*) ... the active masses of reactants participating in a chemical reaction are always proportional to the product of the rate of reaction and the value of K_{eq}.'

(*iii*) ... the active masses of reactants participating in a chemical reaction are proportional to the rate constants for the forward and the reverse reaction.'

(*iv*) ... the rate of reaction is proportional to the active concentration of reacting substances each raised to the power of the index appearing in the balanced stoichiometric equation.'

[The terms 'active mass', 'activity', and 'active concentration' may be regarded as synonymous].

SAQ 1.2b

So far we have been using a standard hypothetical reaction to illustrate the development of the ideas. It is of course necessary to have a scheme which can accommodate differing stoichiometries. This is done by recalling the relationship between the rate and the molecularity of a reaction.

eg $$A + B \underset{(2)}{\overset{(1)}{\rightleftharpoons}} 2C + D$$

$$v_1 = k_1[A][B]$$

$$v_2 = k_2[C][C][D]$$

$$\therefore \quad K_{eq} = \frac{k_1}{k_2} = \frac{[C]^2[D]}{[A][B]}$$

since at equilibrium, $v_1 = v_2$.

Note that for the reaction $A + B \rightleftharpoons C + D$, the concentration units will cancel out in the expression for K_{eq}, ie K_{eq} here is dimensionless. However in the example above, there is only partial cancellation so the units of K_{eq} for this reaction are mol dm^{-3}, if this is the unit of concentration.

SAQ 1.2c

Derive expressions for K_{eq} for the reactions of stoichiometry (*i*) and (*ii*) below.

(*i*) $3A + B \rightleftharpoons 2C + 2D$

(*ii*) $A \rightleftharpoons 2B$

You will see from the last SAQ that it is possible to write a general expression for K_{eq}. If we use *w x*, *y* and *z* for the molecularity of species A, B, C, and D we find that for the reaction:

$$wA + xB \rightleftharpoons yC + zD$$

$$K_{eq} = \frac{[C]^y[D]^z}{[A]^w[B]^x} \text{ at constant temperature and high dilution}$$

These concepts are equally applicable to gas-phase equilibria and to equilibria in solution. In the gas phase, the partial pressures of components are equivalent to the molar concentrations used in solution studies. It is conventional to use the symbol K_p for gas-phase equilibria and K_{eq} for solutions. Some texts use simply K and others use K_c in place of our K_{eq}. Later in this section we shall meet different types of 'K', such as K_a, K_b and K_w. These are standard in all texts; they will be explained as they are introduced.

We shall mainly be concerned with K_{eq} rather than K_p as gas-phase equilibria do not occupy an important place in chemical analysis. However note that K_p and K_{eq} are not numerically the same.

K_p is given by $p_C.p_D/p_A.p_B$, where p = partial pressure. This is because the molar *concentration* of a gas is given by P/RT. Hence the 'RT' part relates K_p and K_{eq}. The actual form of this depends on reaction stoichiometry and need not concern us here. (R is the molar gas constant, and T the temperature in Kelvins).

SAQ 1.2d

Write the equilibrium constant expression, K_{eq}, for the equilibria represented by the equations below.

For each example give also the units of K_{eq}.

(*i*) $PCl3 + P(OEt)3 \rightleftharpoons PCl2(OEt) + P(OEt)2Cl$

(*ii*) $N_2O_{4(g)} \rightleftharpoons 2\,NO_{2(g)}$

(*iii*) $HCOOH_{(aq)} + H_2O_{(l)} \rightleftharpoons H_3O^+_{(aq)} + HCOO^-{}_{(aq)}$

(*iv*) $N_{2(g)} + 3\,H_{2(g)} \rightleftharpoons 2\,NH_{3(g)}$

SAQ 1.2d

SAQ 1.2e

Explain why K_{eq} can never take negative values.

1.2.3. The Position of the Equilibrium

The equilibrium constant, K_{eq}, determines the position of the equilibrium; we might say that it gives an indication of 'how far the reaction has gone'. It is logical at this stage to enquire about the factors which affect the position of the equilibrium. If we perturb an equilibrium system it acquires a non-equilibrium state. Concentrations of the components will change until a new position of equilibrium is established. We described this in a rough-and-ready way by the analogy of the tight-rope walker. In the next few paragraphs we shall develop a useful but scientific rough and ready way of predicting the changes. Consideration of the factors which entered into our definition of K_{eq} will guide us in identifying phenomena which are likely to do this. These factors are of course, concentrations and rate constants for the forward and the reverse reaction. Thus we should be able to predict that changes in temperature and in concentration (or pressure for gas-phase reactions) will influence the position of the equilibrium.

SAQ 1.2f

The dissolution of silver sulphide in pure water involves the following equilibria.

$$Ag_2S_{(s)} \rightleftharpoons 2\,Ag^+ + S^{2-} \qquad (1)$$

$$S^{2-} + H_2O \rightleftharpoons HS^- + OH^- \qquad (2)$$

$$HS^- + H_2O \rightleftharpoons H_2S + OH^- \qquad (3)$$

$$2\,H_2O \rightleftharpoons H_3O^+ + OH^- \qquad (4)$$

Does the fact that the equilibrium constant for equation (1) is *ca.* 6×10^{-50} necessarily mean that silver sulphide will be insoluble in water?

What would be the effect of adding acid to the system?

SAQ 1.2f

SAQ 1.2g

(*i*) If you are informed that the equilibrium constants for the reactions in water of ethylenediaminetetra-acetic acid, EDTA, with most divalent metals are above 10^{10}, does this mean that

(*a*) the complexes formed are very soluble,

(*b*) the equilibrium strongly favours the formation of the complex,

or

(*c*) the complexes are formed at pH values higher than 10?

(*ii*) What information is conveyed by the qualitative statement that the equilibrium constants for the dissociation of both aqueous ammonia and aqueous formic acid (methanoic acid) are moderately low?

SAQ 1.2g

1.2.4. The Theorem of Le Chatelier and Van't Hoff

This theorem was published in 1895 and is frequently called simply Le Chatelier's principle. The principle has been criticised on theoretical grounds but it remains a most useful 'rule of thumb' for predicting the direction in which the *position* of equilibrium will change as other factors are altered. The theorem may be stated as follows: 'Any alteration in the factors which determine the position of a chemical equilibrium leads to a displacement of the position of equilibrium in such a way as to *oppose* the effect of the alteration.'

Thus an increase in temperature will lead to an equilibrium shifting in the direction which will *absorb* heat. An increase in pressure will similarly lead to a shift in the direction which will lead to a reduction in volume. Some examples follow.

(*a*) The dissolution of many solids is endothermic. If we consider a solid in equilibrium with its saturated solution, the system will respond to an increase in temperature by allowing more solid to dissolve.

(*b*) The neutralisation of H_3O^+ by OH^- to form water is strongly exothermic, hence an increase in temperature leads to a greater degree of dissociation of water into H_3O^+ and OH^-.

(*c*) The equilibrium which features in the well-known Haber process for the manufacture of ammonia involves four volumes of gas on the left-hand side (as written below) and two volumes on the right-hand side. Consequently an increase in pressure shifts the equilibrium to the right-hand side, ie to the side with the smaller volume.

$$N_2 + 3H_2 \rightleftharpoons 2NH_3$$

(*d*) The equilibrium system of equimolar quantities of iron(III) and thiocyanate ion lies fairly strongly to the right-hand side and the complex formed has a deep blood-red colour.

$$Fe^{3+} + SCN^- \rightleftharpoons (FeSCN)^{2+}$$

The addition of chloride ions (as ammonium chloride) effectively removes iron by the following reaction.

$$Fe^{3+} + 4Cl^- \rightleftharpoons (FeCl_4)^-$$

Le Chatelier's theorem predicts a shift in the thiocyanate equilibrium to the left-hand side as the system tries to replace the iron removed as $(FeCl_4)^-$. This is confirmed in practice by the observation of a diminution in the depth of the red colour.

SAQ 1.2h

Apply Le Chatelier's principle to the following.

(*i*) What will be the effect of an increase in temperature on the following equilibrium?

$$H_2 + I_2 \rightleftharpoons 2HI$$

$$\Delta H^o = +9.6 \text{ kJ mol}^{-1}$$

⟶

SAQ 1.2h (cont.)

(*ii*) What will be the effect of an increase in pressure on the following equilibria?

(*a*) $SO_3 + NO \rightleftharpoons SO_2 + NO_2$

(*b*) $2NO_2 \rightleftharpoons 2NO + O_2$

(*iii*) A solution of hydrogen iodide, HI, in aerated water rapidly becomes browner as the temperature is raised. Is the reaction of HI with oxygen exothermic or endothermic?

(*iv*) Does the observation that the reaction of the oxalate ion, $(COO)_2^{2-}$, and the permanganate ion, MnO_4^-, proceeds more rapidly when the temperature is raised necessarily indicate that the reaction is endothermic?

1.3. THE EQUILIBRIUM CONCEPT IN ANALYTICAL SCIENCE

At this stage we have developed a useful general model of equilibrium systems. We recognise that a wide variety of chemical systems are properly regarded as being in equilibrium and consequently are involved in dynamic exchange, even when reaction appears to have ceased. The same goes for chemical processes which involve phase changes or dissolution in a solvent.

Clearly, the analytical chemist is concerned with the collection of samples, their pre-treatment and/or concentration, and the determination of the analyte. Equilibrium systems may be relevant at all these stages in the totality of a chemical analysis.Indeed, chemists will readily acknowledge the role of equilibrium studies in classical wet analysis, involving techniques such as precipitation and titration. However, because of the inclusion of both collection and pre-treatment as part of the analytical process, equilibria are also of interest for instrumental methods of analysis.

Some of the topics for which equilibria are important are listed below.

Collection of gases,
Oxidation-reduction,
Indicators,
Neutralisation,
Variable valency,
Extraction into non-aqueous phases,
Derivatisation,
Dissociation of water,
Buffers, pH, and buffer action,
Weak and strong electrolytes,
Hydrolysis of salts,
Low solubility compounds
The common-ion effect,
Complexes,
Stepwise equilibria.

You may have met some of these in your previous studies, others you will be meeting for the first time. We shall treat these features in the all-embracing context of equilibria and analysis, rather than that of a detailed study of the phenomena in their own right.

The way in which our equilibrium model is important for the application of these topics in analytical chemistry is indicated in the next Section. In the more detailed approach which follows, ideas are examined qualitatively, that is we take a rough-and-ready look at the way the systems behave. This is followed by a quantitative approach, often involving calculations. It is usually good learning strategy to make sure that you understand the qualitative part before plunging into the quantitative section.

1.4. APPLICATIONS OF EQUILIBRIA

An Example of Multiple Equilibria

One of the recommended methods for the determination of sulphur dioxide in smoke, in air-pollution studies, involves drawing the sample through a solution of hydrogen peroxide by using a fritted bubbler. The sulphur dioxide is oxidised to sulphuric acid and this is then titrated with a base, with phenolphthalein as the indicator.

Π How many relevant equilibria can you identify?

You should have identified at least two; if you got more, then well done. There are in fact four relevant equilibria here, and if you count the ionisation of the solvent there are five.

First the gas dissolves in the water; clearly we have a dissolution equilibrium here. This is very relevant to our assumptions about the efficiency of the collection, for we must assume that the gas is very soluble, ie the dissolution equilibrium lies very much over to the right-hand side, for successful collection by bubblers.

Then there is the oxidation. As we are planning to titrate the sulphuric acid rather than to determine sulphur dioxide directly (eg by ir absorption in the gas phase), we must assume a quantitative conversion. That is the equilibrium for the reaction must lie very strongly over to the right-hand side.

The third and fourth equilibria involve the reactions which take place during the titration, ie the acid–base equilibrium and the indicator equilibrium. The details of these are developed later in this section.

Variable Valency

A similar interest in equilibria will arise if separate determinations of a metal in two oxidation states are required, eg Cr(III) and Cr(VI). We clearly need to recognise that the equilibrium which operates in the medium prior to collection may be grossly disturbed by the act of collection and by subsequent handling during determination. Can we assume that the ratio of Cr(III) to Cr(VI) in our laboratory sample is the same as that present in the stream from which the sample came?

Concentration

One of the techniques used for the concentration of samples for trace analysis may involve several extractions of a solute from an aqueous phase into a non-aqueous phase such as trichloromethane (chloroform). In this way a substance which is present at such low levels that it would evade detection may be sufficiently concentrated to allow analytical measurement. Each extraction of course involves an equilibrium process.

Derivatisation

In yet other methods an analyte may be derivatised, ie a reaction is carried out to convert the analyte into a chemical derivative which is much more readily determinable than the original analyte. We can see that some knowledge of the equilibrium nature of the derivatisation process itself is crucial to the extrapolation of measurements on the derivative back to a statement about concentrations of the original material.

Modification of Standard Methods

Practical instructions in analytical chemistry often have a rather 'recipe-like' appearance; these are the prescribed or standard methods. The equilibrium concept is necessary for an understanding of why many of these methods take the form they do. However on occasions you may wish, quite legitimately, to modify a standard method to suit your own local needs. Again, a sound working knowledge of equilibrium processes is crucial if modifications are to be valid and acceptable.

The Dissociation of Water, pH, and Buffer Action

Water is an ubiquitous solvent in most branches of chemical analysis and it should come as no surprise that the equilibrium behaviour of water becomes involved in many other equilibria.

Water undergoes self-dissociation according to the equation below.

$$2\,H_2O \rightleftharpoons H_3O^+ + OH^-$$

You will recall that the pH of a system is determined by the hydrogen-ion concentration. We can clearly see that the acid–base behaviour of a system is intimately connected with this equilibrium. Further we may note that the purpose of buffer systems is to maintain a constant value of pH. Equilibrium theory allows us to understand buffer action.

Indicators

The action of indicators, whether for neutralisation, complexometric, redox, or precipitation titrations, must be reversible. This must be so if for no other reason than to accommodate local high concentrations as reagents are mixed. Thus indicator systems are all equilibrium systems and a clear view of the relationship(s) between these and other equilibrium systems operating during the analysis is important, if the analytical results produced are to be quantitatively meaningful.

Dissociation of Weak Electrolytes and the Hydrolysis of Salt

Weak acids and weak bases are those for which the equilibrium does not lie strongly over to the righthand side of the dissociation equation. You will recall that this is equivalent to saying that they are only weakly dissociated. The behaviour of such systems may be particularly sensitive to the presence of other substances, particularly those which affect acid--base behaviour. Thus a knowledge of equilibria is essential to an understanding of both their determination as analytes, as well as their potential participation in buffer systems.

By the same token, salts of weak acids and weak bases may undergo hydrolysis in aqueous solution, and a knowledge of appropriate equilibrium constants permits the calculation of the pH of such systems.

Low Solubility Compounds and the Common-ion Effect

The analyst frequently encounters the phenomenon of precipitation, ie the formation *in situ* of a compound of very low solubility. This may be part of a procedure for removing unwanted material, for concentrating a fraction which contains a desired analyte, or for determining the analyte itself. We are clearly interested in ensuring that only an absolute minimum of the material remains in the solution and it is desirable to be able to quantify this. Furthermore we wish to be confident that the precipitation has produced a pure material uncontaminated with other substances.

The minute solubilities with which we are concerned may be described as arising from equilibrium systems. We shall be mainly interested in solute (solid)–solute (solution) equilibria and any equilibria dealing with subsequent dissociation of the solute in solution. Knowledge of equilibrium systems also suggests that these low solubilities will be influenced by the presence of materials with a common anion or cation, the so-called 'common-ion' effect.

The Formation of Complexes and Stepwise Equilibria

Coordination compounds, or complexes, occur in most branches of chemical analysis, and they have a wide diversity of roles. Some of the most common are agents to aid detection and measurement in visible spectroscopy, agents to aid extraction into non-aqueous solvents during pre-treatment, masking agents for the effective removal of certain metals from the system, indicator systems, and the analytical reagents themselves in complexometric titration.

Coordination equilibria are important for all these, and a knowledge of equilibrium constants is necessary for the proper application of coordination chemistry to chemical analysis. Also, as some complexes involve stepwise addition of more than one ligand (coordinating group), we must concern ourselves with the related stepwise equilibria.

Redox Equilibria

Many analytical methods exploit the fact that some elements exist in several oxidation states. The selection of appropriate reagents and appropriate indicators for redox processes requires consideration of these redox reactions as equilibria.

Objectives

On completion of this Part of the Unit, you should now be able to:

- make accurate qualitative statements about equilibrium systems;

- recognise the operational limitations of equilibrium concepts;
- relate equilibrium concepts to a range of procedures in chemical analysis;
- do calculations relevant to the application of equilibrium concepts to chemical analysis.

2. Solution Phenomena

This Part examines the nature of the dissolution of substances in water and the effect of such dissolution on the ionisation of water. Models for acids and bases are briefly examined and the concept of strong and weak electrolytes is developed. The equilibrium model is used to describe the behaviour of both strong and weak acids and bases during titrations; also the effect of the presence of ions common to the reacting compounds. These concepts are developed further to explain the behaviour of buffer systems and the hydrolysis of salts.

2.1. DISSOLUTION AND THE EQUILIBRIUM PROCESS

The dissolution of a solute (which may be a solid, liquid, or gas) in a solvent is generally described at a molecular level in terms of an interaction of the solvent molecules with solute species. This is followed by their transport as solvated species from the interface into the bulk solvent. This sort of model for dissolution does not consider the changes undergone by individual solvent molecules, but in fact there is usually very rapid exchange of solvent molecules between those doing the actual solvating and those in the bulk of the solvent. However, for some solvents this type of solvation does not represent the limit of their participation in solution phenomena. Solvents which are themselves weak electrolytes may undergo a process known as *Auto-Ionisation* or *Self-Ionisation*, that is they are weakly dissociated themselves.

eg

$$2H_2O \rightleftharpoons H_3O^+ + OH^-$$

$$2CH_3COOH \rightleftharpoons CH_3COOH_2^+ + CH_3COO^-$$

$$2NH_3 \rightleftharpoons NH_4^+ + NH_2^-$$

You will see from the above that this process is in fact simply proton transfer from one solvent molecule to another. The process is then called *Autoprotolysis*.

SAQ 2.1a

Which of the following represent autoprotolysis?

(*i*) $HF + HF + \rightleftharpoons H_2F^+ + F^-$

(*ii*) $2H_2SO_4 \rightleftharpoons H_3SO_4^+ + HSO_4^-$

(*iii*) $2H_2SO_4 \rightleftharpoons H_2SO_3 + HSO_4^- + OH^-$

(*iv*) $HF + HF \rightleftharpoons HF_2^- + H^+$

(*v*) $2SO_2 \rightleftharpoons SO_3^{2-} + SO^{2+}$

(*vi*) $2HCN \rightleftharpoons H_2CN^+ + CN^-$

You will note that for water, auto-ionisation has produced both H_3O^+ (often written simply as H^+ for convenience), and OH^-. These ions are, of course, generally associated with acidic and alkaline situations respectively.

The Arrhenius classification of acids and bases which was developed in the late nineteenth century defined acids as those hydrogen-containing compounds which produce H^+ when dissolved in water. Likewise bases were those compounds which release OH^- when dissolved in water. Thus if we look back at our dissociation of solvents we can see that water would be described as both an acid and as a base in this model.

The role of solvent dissociation was more fully developed in the treatment of acids and bases by Brõnsted and Lowry (1923). In the Broñsted–Lowry view an acid is a compound capable of donating a proton and a base simply one capable of accepting a proton.

SAQ 2.1b

Mark the following as 'true' or 'false', within the context of the Brõnsted–Lowry model.

		True	False
(*i*)	Electron donors are known as acids.		
(*ii*)	Acids react with proton donors.		
(*iii*)	Acids are themselves proton donors.		
(*iv*)	Water auto-ionises to produce H^+ therefore never acts as a base.		
(*v*)	Acids are proton acceptors.		
(*vi*)	Bases are proton acceptors.		

SAQ 2.1b

If an acid is a proton donor we can show it donating a proton to water.

Thus we can write:

$$H_2O + \text{'acid'} \rightleftharpoons \text{'anion'} + H_3O^+$$

However we can see that this is an equilibrium process and that it is possible to write it thus.

$$H_3O^+ + \text{'anion'} \rightleftharpoons H_2O + \text{'acid'} \qquad (2.1)$$

But recall that the Brönsted–Lowry definition of a base is 'a compound capable of accepting a proton', ie the 'anion' here is behaving as a base. We normally describe this as the conjugate base of the parent acid.

$$H_2O + \text{parent acid} \rightleftharpoons \text{conjugate base} + H_3O^+$$

Π How would you describe the role of Cl^- in the equilibrium below?

$$H_2O + HCl \rightleftharpoons H_3O^+ + Cl^-$$

Here Cl^- is formed by removing a proton from HCl, ie Cl^- would accept a proton to become HCl. Thus it is the conjugate base of the parent acid HCl.

We may now examine the behaviour of bases with the confident expectation of something similar.

Π When ammonia is dissolved in water the following reaction takes place.

$$H_2O + NH_3 \rightleftharpoons NH_4^+ + OH^- \qquad (2.2)$$

Which is the parent base and which the conjugate acid for ammonia?

Again look at the relationship in terms of 'which species gets a proton and which species loses one'. Here NH_3 gains a proton so ammonia is the parent base of the conjugate acid NH_4^+.

SAQ 2.1c

(1) Give the formula of the conjugate base of each of the following acids:

(*a*) C_6H_5COOH,

(*b*) HCO_3^-,

(*c*) H_2CO_3,

(*d*) NH_4^+.

(2) Give the formula of the conjugate acid of each of the following bases:

(*a*) HPO_4^{2-},

(*b*) Cl^-,

(*c*) NH_3,

(*d*) OH^-.

SAQ 2.1c

Π Look at Eq. (2.1) again, but this time consider the behaviour of the species H_2O and H_3O^+ in the light of the Brönsted–Lowry definition of acids and bases. Are the species proton acceptors or proton donors?

In Eq. (2.1) H_2O is clearly a proton acceptor therefore H_2O is behaving as a base. At the same time H_3O^+ is behaving as a proton donor, thus H_3O^+ is an acid.

For ease of identification we shall refer to these as base(2) and acid(2). The original parent acid and its conjugate base will be referred to as acid(1) and base(1).

Thus we can write Eq. (2.1) as:

$$\text{base(2)} + \text{acid(1)} \rightleftharpoons \text{acid(2)} + \text{base(1)}$$

Π Convince yourself that both the following statements are true:

H_3O^+ is the conjugate acid of parent base H_2O; H_2O is the conjugate base of the acid H_3O^+.

Examine Eq. (2.2) and identify the function of H_2O and OH^-.

The hydroxide ion is naturally acting as a proton acceptor and is thus a base, ie base(2). Here H_2O provides the protons so H_2O is acting as an acid, ie acid(2).

Eq. (2.2) thus becomes:

H_2O +	NH_3	$\rightleftharpoons$	NH_4^+	+ OH^-
acid(2)	parent base ie base(1)		conjugate acid ie acid(1)	base(2)

The important consequence of this is that neutralisation is to be seen as the operation of the following general equilibrium.

$$\text{acid(1)} + \text{base(2)} \rightleftharpoons \text{acid(2)} + \text{base(1)}$$

This is an equilibrium process and consequently could move in either direction. In practice, the reaction spontaneously moves in the direction which will produce a predominance of the least dissociated acid–base pair, ie the weaker acid and base.

A further consequence of this model of acid–base behaviour is the treatment of the dissolution of many solutes as neutralisation reactions. A solvent may behave principally as an acid or principally as a base depending on the nature of the solute. We notice however that the ubiquitous solvent water is able to function as both, ie it is *amphiprotic*. These ideas will be developed later in the text.

SAQ 2.1d

(1) Which of the following statements are correct?

(*a*) H_3O^+ is the conjugate acid of H_2O.

(*b*) H_3O^+ is the conjugate acid of OH^-.

(*c*) H_3O^+ is the conjugate base of OH^-.

(*d*) OH^- is the conjugate base of H_3O^+.

(*e*) OH^- is the conjugate base of H_2O.

(*f*) H_2O is the conjugate base of H_3O^+.

(*g*) H_2O is the conjugate acid of OH^-.

(2) For the following equations arrange the materials in the form;

$$\text{acid}(1) + \text{base}(2) \rightleftharpoons \text{acid}(2) + \text{base}(1)$$

(*a*) $HCl + H_2O \rightleftharpoons H_3O^+ + Cl^-$

(*b*) $H_2O + CH_3CH_2COOH \rightleftharpoons H_3O^+ + CH_3CH_2COO^-$

(*c*) $H_2O + CO_3^{2-} \rightleftharpoons HCO_3^- + OH^-$

(*d*) $Al(H_2O)_4(OH)_2^+ + H_2O \rightleftharpoons H_3O^+ + Al(H_2O)_3(OH)_3$

(*e*) $CH_3COO^- + H_3O^+ \rightleftharpoons CH_3COOH + H_2O$

(*f*) $H_2O + Al(H_2O)_6^{3+} \rightleftharpoons AlOH(H_2O)_5^{2+} + H_3O^+$

(*g*) $CH_3NH_2 + H_2O \rightleftharpoons OH^- + CH_3NH_3^+$

SAQ 2.1d

The approach to use here, if you had any difficulty, is to identify each acid–base pair, first without reference to which is which, eg HCl/Cl^-, $CH_3NH_3^+/CH_3NH_2$. Then decide which is the acid and which is the base as in the previous example. You can treat the role of water similarly, ie H_3O^+/H_2O is an acid–base pair as is H_2O/OH^-.

2.2. THE IONIC PRODUCT FOR WATER, K_w

If we return to our solvent system of water undergoing dissociation into H_3O^+ and OH^- we may represent this as the equilibrium given below.

$$2\,H_2O \rightleftharpoons H_3O^+ + OH^-$$

Consequently we may apply an equilibrium constant K_{eq} to this process.

$$K_{eq} = \frac{[H_3O^+][OH^-]}{[H_2O]^2}$$

The concentration of the solvent itself may undergo minor changes because of the dissociation of the solute in it, but to all intents and purposes the concentration of the solvent is constant, at any fixed temperature. (As the density of the solvent changes with temperature the concentration of the solvent would also change). Thus at constant temperature the quantity $[H_2O]^2$ is also constant and we can use a new term K_w to replace the product of K_{eq} and $[H_2O]^2$.

$$K_w = [H_3O^+][OH^-]$$

We see that K_w is simply the *product* of the concentrations of the ions formed by the auto-ionisation of water; it is known as the *ionic product for water*. K_w has a value of approximately 10^{-14} mole^2 dm^{-6} at 25 °C, and we can see that in neutral water for which $[H_3O^+] = [OH^-]$ the value of these will *both* be 10^{-7} mole dm^{-3}. Note that these values will change with changing temperature for two reasons, first a shift in the fundamental equilibrium, and secondly a change in the density of water with consequent change in its concentration. In fact K_w ranges from 0.11×10^{-14} at 0 °C to 5×10^{-14} at 100 °C (units are mol^2 dm^{-6}).

SAQ 2.2a

Given that $K_w = 1 \times 10^{-14}$ mol^2 dm^{-6} at 25 ° C, does this mean that the concentration of H^+ ions in pure water is (*i*) $10^{-14}M$, (*ii*) $10^{-7}M$, (*iii*) $10^{-28}M$, or (*iv*) $10^{+14}M$?

Π By using a value for K_w of 1.00×10^{-14} mol^2 dm^{-6}, calculate the concentration of OH^- ions at the following pH values. Plot a graph of $[H^+]$ versus $[OH^-]$ using a scale factor of 10^7. Comment on the relationship between the rectangular area under the curve and K_w.

(For this exercise recall that $\text{pH} = -\log[H^+]$).

Carry out the calculation for the following pH values:

6.50, 6.60, 6.80, 7.00, 7.20, 7.40, 7.50.

As this involves similar manipulation of the figures for each pH value a tabular approach is most convenient.

Your results should look like this:

pH	6.5	6.75	etc
$10^7[H^+]/M$	3.16	1.78	etc
$10^7[OH^-]/M$	0.316	0.562	etc

When you have plotted your graph, it should have the form of a rectangular hyperbola. One of the properties of this type of function is that the rectangles which can be drawn beneath the curve all have the same area. In our case, this is simply the product of the sides of the rectangles, ie $[H^+] \times [OH^-]$, *viz*, the area corresponding to K_w.

SAQ 2.2b

The *International Critical Tables* give the following values for K_w at different temperatures. Units are all $mol^2\ dm^{-6}$.

T °C	K_w
0	1.17×10^{-15}
50	5.49×10^{-14}
100	5.12×10^{-13}
150	2.34×10^{-12}
200	5.49×10^{-12}

Which of the following are true or false about pure water?

(*i*) $[H^+]$ increases and $[OH^-]$ decreases as the temperature increases.

(*ii*) $[H^+]$ decreases and $[OH^-]$ increases as the temperature increases.

(*iii*) Both $[H^+]$ and $[OH^-]$ increase as the temperature increases. ⟶

SAQ 2.2b (cont.)

> (*iv*) $[H^+]$ and $[OH^-]$ remain constant as the temperature increases.
>
> (*v*) $[H^+]$ and $[OH^-]$ decrease as the temperature increases.

2.3. ACID AND BASE DISSOCIATION

It is usually sufficient to use either K or K_{eq} to indicate an equilibrium constant; some texts will use K, others will use K_{eq}. We have already introduced a special constant, K_w, the ionic product for the dissociation of water, and here we develop the idea to include the dissociation of acids and bases. To do this we use the associated constants K_a and K_b.

For the dissociation of acids we can write:

$$HA + H_2O \rightleftharpoons H_3O^+ + A^-$$

$$\therefore \qquad K_{eq} = \frac{[H_3O^+][A^-]}{[HA][H_2O]}$$

Again, as we did for pure water, we assume that the concentration of water remains constant. Similarly, we define K_a as K_{eq} $[H_2O]$.

$$\text{Hence} \qquad K_a = \frac{[H_3O^+][A^-]}{[HA]} \quad \text{mol dm}^{-3}$$

Π Recall the definition of a base and write out the equation for the dissociation of water induced by a base, B. Hence deduce the expression for the equilibrium constant, and by replacing $K_{eq} \times [H_2O]$ by K_b, obtain an expression for K_b.

This is, of course, similar to the dissociation of water induced by acids, so for the base B we have;

$$H_2O + B \rightleftharpoons BH^+ + OH^-$$

and $$K_{eq} = \frac{[BH^+][OH^-]}{[H_2O][B]}$$

Now $$K_b = K_{eq}[H_2O]$$

$$\therefore \quad K_b = \frac{[BH^+][OH^-]}{[B]} \quad \text{mol dm}^{-3}$$

The Relationship Between K_w, K_a and K_b

If we take a weakly dissociated acid, HA and examine the equation for its dissociation, we note that the anion, A^- is the conjugate base of HA. The constant K_a refers to this equilibrium.

We may also quite reasonably examine the reaction we might expect between this base, A^- and water and derive the related expression for K_b for the base A^-.

Π Derive an expression for K_b for the base A^-. Refer to the expression for K_a for the acid, HA, given above, and examine the result of multiplying K_a by K_b.

Starting with the equation for the reaction of A^- as a base, with water, we produce the conjugate acid HA.

$$A^- + H_2O \rightleftharpoons HA + OH^-$$

Hence we can write

$$K_b = \frac{[HA][OH^-]}{[A^-]}$$

(You could of course have obtained this simply by substituting A^- for B and recalling that $A^- + H^+ = HA$).

Looking at this expression for K_b and the equation and for K_a,

$$K_a = \frac{[H_3O][A^-]}{[HA]}$$

notice that one expression contains the concentration of hydronium ion $[H_3O^+]$ and the other contains the concentration of hydroxide ions $[OH^-]$. If we multiply the expression for K_a by that for K_b of the conjugate base we obtain simply the product $[H_3O^+][OH^-]$, ie K_w.

$$K_a \times K_b = \frac{[H_3O^+][A^-]}{[HA]} \times \frac{[HA][OH^-]}{[A^-]}$$

$$= [H_3O^+][OH^-] = K_w$$

ie $$K_w = K_a \times K_b$$

This sort of expression is practically useful only when the compound HA is neither strongly dissociated nor practically undissociated; ie the equilibrium lies neither strongly to the r.h.s. nor to the l.h.s. of the dissociation equation.

You will of course quite reasonably ask, what constitutes high and low values of K_a and K_b. Recall that K_w has a value of *ca* 10^{-14}, and you can see that the actual numbers involved are very small indeed by every day standards. It also turns out that the values themselves vary through several orders of magnitude, ie several powers of 10. It is in practice convenient to introduce a new term pK, in addition to our K values. You will recall from previous study that hydrogen-ion concentration is frequently referred to by means of the term pH, pH being given by $-\log_{10}[H^+]$. The pK terms are defined in an analogous manner.

$$pK_a = -\log_{10} K_a$$

$$pK_b = -\log_{10} K_b$$

This leads to a useful range of pK_a and pK_b values from about 1 to about 5 in normal circumstances. There are of course many compounds, which we would describe as minutely dissociated, with pK_a and pK_b values outside this range.

2.4. STRONG AND WEAK ELECTROLYTES

You will recall from previous studies that the term *strong electrolyte* refers to those compounds which are almost completely dissociated when dissolved in water. *Weak electrolytes* are, of course, compounds which are only partially dissociated under similar conditions. These are very much qualitative terms, that is they are not intended to indicate with any degree of precision just what the degree of dissociation actually is.

With the equilibrium model we have developed can you say whether K_{eq} will be high or low for weak electrolytes? In an equilibrium context, we can say that for strong electrolytes the equilibrium constants for dissociation will be very large, for weak electrolytes they will be only moderate to small. You should note that some texts use the term *electrolyte* alone (ie without the qualifier 'strong') to imply a strong electrolyte. You can see where the word comes from if you remember that dissociation produces ions and that ions can carry an electric current; electrolytes are essentially 'things which can be electrolysed', ie decomposed by the passage of an electric current.

Many common inorganic acids (often called 'mineral acids') such as nitric acid, hydrochloric acid, and hydrobromic acid are strong electrolytes in water. Most soluble salts and soluble hydroxides of elements of group I and II of the periodic table are also strong electrolytes.Note however, that it does not necessarily follow that because a compound is strongly dissociated in water it will be in other solvents.

In contrast to the behaviour described above, the feature of interest for weak electrolytes is that when dissolved in water the solution contains a significant concentration of both ions (from the dissociation) *and* the undissociated molecules themselves.Many organic acids and organic bases fall into this category (notable exceptions are some sulphonic acids and some chlorinated derivatives of ethanoic acid). The so-called weak inorganic acids such as boric acid,carbonic acid, and phosphoric acid as well as some slightly soluble hydroxides are also weak electrolytes.

Some examples of K_a, pK_a, K_b and pK_b values for aqueous solutions are given in Fig. 2.4a.

	K_a	pK_a
p-Aminobenzoic acid	1.2×10^{-5}	4.92
Benzoic acid	6.46×10^{-5}	4.19
Benzenesulphonic acid	2×10^{-1}	0.70
2,2-Dichloroacetic acid	3.32×10^{-2}	1.48
Glycerol (1,2,3-Trihydroxy propane)	7×10^{-15}	14.15
Naphthalenesulphonic acid	2.7×10^{-1}	0.57
Deuterioacetic acid (in D_2O)	5.5×10^{-6}	5.26
Sulphuric acid	1.2×10^{-2}	1.92
Nitrous acid	4.6×10^{-4}	3.34 (12.5 °C)
Lactic acid (2-Hydroxy propanoic acid)	1.4×10^{-4}	3.85
Phenol	1.0×10^{-10}	10.0
Histidine	6.7×10^{-10}	9.17
	K_b	pK_b
Ammonia solution (aqueous)	1.77×10^{-5}	4.75
Zinc hydroxide	9.6×10^{-4}	3.02
Beryllium hydroxide	5×10^{-11}	10.30
Aniline	4.27×10^{-10}	9.37
Pyridine	1.48×10^{-9}	8.83
Hydroxylamine	1.07×10^{-8}	7.97

Fig. 2.4a. *Examples of K_a, pK_a, K_b and pK_b values for aqueous solutions (mainly at 25 °C)*

SAQ 2.4a

Look at the table of pK_a and pK_b values above, and remembering that the following classifications are rather broad, classify the compounds as strong, weak, or very weak electrolytes.

2.5. NEUTRALISATION TITRATIONS

You will recall that in neutralisation titrations we are carrying out an acid–base reaction. The essential feature is the addition of one component about which we have all relevant information (volume, concentration, etc) to a solution for which information is partially lacking. We assume the reaction to be quantitative. If the indicator is correctly chosen, the pH at the end-point will correspond to that pertinent to the equivalence point, so that if we know the volume added and the molarity of the titrant, and the volume of the solution of unknown molarity, this last is readily calculated.

You will notice the assumptions here. First, the assumption about a quantitative reaction (ie the reaction goes to completion) implies knowledge of the *equilibrium behaviour* of the system; secondly, we assume that we can relate observations of *pH changes* to the equilibrium behaviour of the system.

In neutralisation titrations it is usual to work with strong acids and strong bases if possible,that is compounds with very low values for pK_a and pK_b. In these circumstances it is reasonable to assume that our compounds are *totally dissociated*; thus the following further simplifying assumptions are possible.

(*a*) For acids the hydrogen-ion concentration is given directly by the molarity of the acid (or double this for strong dibasic acids).

(*b*) For bases the hydroxide-ion concentration is given directly by the molarity of the base (or double this for strong diacid bases).

The hydrogen-ion concentration and the hydroxide-ion concentration are easily related *via* K_w.

SAQ 2.5a

Can you recall the relationship between pH and the hydrogen-ion concentration of a solution?

Try the following, assuming complete dissociation of the solute in water.

(*i*) Calculate the pH of 0.05*M*-HCl.

(*ii*) Calculate the pH of 0.05*M*-NaOH.

SAQ 2.5b

Calculate the hydrogen-ion concentration of the following materials for which pH values are given:

lemon juice, 2.2; beer, 4.5; blood, 7.7; seawater, 8.3.

For strong acid-strong base titrations, equilibrium considerations may be virtually discounted except in the very small region for which $[H^+]$ is approximately equal to $[OH^-]$, ie as the titration moves through the neutralisation-point.

Work through the following SAQs and plot your results as a graph of volume added on the *x*-axis and pH on the *y*-axis.

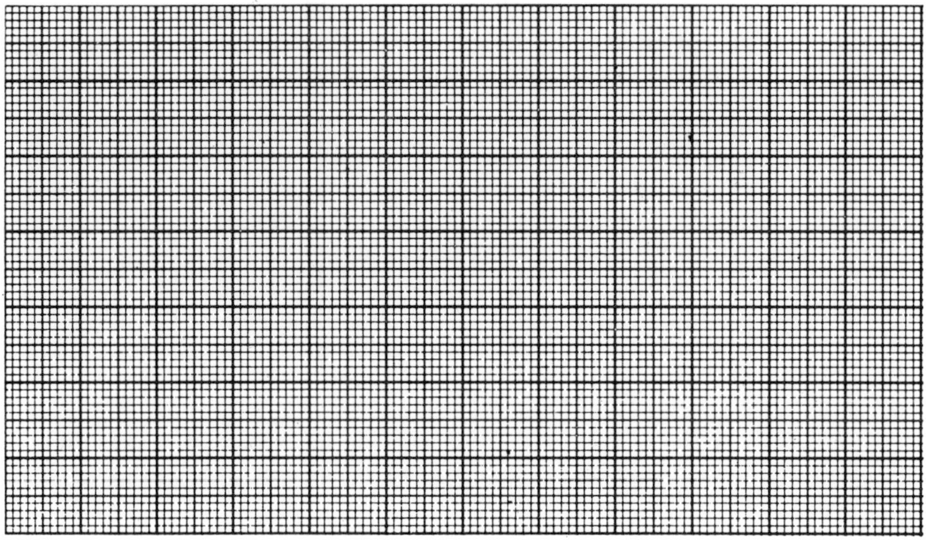

SAQ 2.5c

Calculate the pH at each stage of a titration in which 14.0 cm^3 of 0.50*M*-NaOH is added to 10.0 cm^3 of 0.50*M*-HCl in 2.0 cm^3 portions. Do not forget the dilution which arises from the addition.

SAQ 2.5d

By using a technique similar to that in the previous question, calculate the pH change when the following volumes of 0.01*M*-NaOH are added to 10.0 cm^3 of 0.005*M*-HCl: 0.0, 4.0, 4.5, 4.8, 5.0, 5.2, 5.5, 6.0 cm^3.

2.6. WEAK ACIDS AND WEAK BASES

We have seen that equilibrium considerations do not lead to any particular problems when we consider neutralisation of strongly dissociated acids with strongly dissociated bases. Indeed you will find that the section dealing with neutralisations in detail gives you a certain amount of latitude when selecting experimental conditions for this sort of work.

However there are other situations which we must bear in mind. These are briefly neutralisations of the following types:

(*a*) weak acid–strong base,

(*b*) weak base–strong acid,

(*c*) weak base–weak acid.

For the neutralisation of a weak acid by a strong base it is clear that during the titration we produce a salt (acid plus base gives salt plus water). It is also a distinct possibility that this salt will be a strong electrolyte. The situation we now have is that of a salt which is a strong electrolyte in the same solution as a related acid which is a weak electrolyte.What is more, the salt shares the anion with the anionic part of the acid.

The same reasoning will apply for the titration of a weak base with a strong acid. The product of this reaction will be a strong electrolyte which shares the cation with the cationic part of the base. We refer to this shared ion as the *common ion*, ie it is common to one of the starting materials and the product. If you think about this carefully you will see that there are bound to be common ions in the titration of strong acids with strong bases. However, in these cases the common ion has no impact on the equilibrium behaviour of the system because the reactants themselves are strongly dissociated. We become specifically interested in the common ion when the reactants are not strongly dissociated. The phenomenon is called the *common-ion effect*.

2.7. THE COMMON-ION EFFECT

The question we must address is this: 'What is the effect of the common ion on equilibrium behaviour?' In looking at this question we must recognise that it has a more general context, ie it is not restricted to the conditions which arise during titrations, but should also be considered whenever we have salt–acid or salt–base mixtures. Later in this section we shall use these ideas to study buffer systems.

We can begin by examining the effect qualitatively.

Π What will be the general nature of the effect of the addition of sodium ethanoate to a solution of ethanoic acid (sodium acetate to acetic acid)?

Start by examining the expression for the dissociation of ethanoic acid and rearrange it to give the hydrogen-ion concentration.

$$K_a = \frac{[H^+][Ac^-]}{[HAc]} \quad (Ac^- = \text{ethanoate})$$

$$[H^+] = \frac{K_a[HAc]}{[Ac^-]}$$

We can see that the addition of sodium ethanoate will lead to an increase in the concentration of the ethanoate ion, $[Ac^-]$ thus leading to a decrease in the hydrogen ion concentration, ie an increase in pH. In other words the addition of the common ion, here the ethanoate ion, has *decreased the degree of ionisation* of the acid (defined later).

But what sort of dissociation do we expect for a weak acid anyway? We know from earlier parts of this section that the dissociation (or ionisation) is expected to be very small. Try to develop the idea by calculating the pH of the weak acid given below, by assuming that the dissociation is very slight, ie we can assume that the concentration of HA is not significantly altered by the dissociation.

Π What is the pH of a solution of $0.5M$-formic acid (methanoic acid)? ($K_a = 1.77 \times 10^{-4}$).

Start by writing out the dissociation equilibrium, then the expression for K_a in terms of concentrations.

$$HCOOH \rightleftharpoons H^+ + HCOO^-;$$

$$K_a = \frac{[H^+][HCOO^-]}{[HCOOH]}$$

The stoichiometry requires equal numbers of both ions to be produced whatever the value of K_a.

Hence $$[H^+] = [HCOO^-]$$

Therefore we can rewrite the expression for K_a as:

$$K_a = \frac{[H^+]^2}{[HCOOH]}$$

Because K_a is quite small we can assume that the concentration of formic acid present in solution as the undissociated acid is approximately 0.50 mol dm^{-3}

ie $$[HCOOH] \gg [HCOO^-]$$

$$\therefore \quad [HCOOH] = ca\ 0.50 \text{ mol dm}^{-3}$$

Thus $$[H^+]^2 = K_a \times [HCOOH]$$

$$[H^+] = [0.88_5 \times 10^{-4}]^{\frac{1}{2}}$$

$$pH = -\log [H^+] = 2.03$$

In order to pursue this quantitative line of reasoning, it is quite useful to introduce a term for the degree of ionisation, *viz* α, and relate this to K_a. Our eventual purpose in doing this is to enable us to relate variations in α to changes in concentration of the common ion by means of the equilibrium concept.

Let us recall our standard dissociation equilibrium.

$$HA \rightleftharpoons H^+ + A^- \quad ; \quad K_a = \frac{[H^+][A^-]}{[HA]}$$

We now define the degree of ionisation α as the fraction of HA dissociated, ie α will be between 0 and 1.

At this stage we have to be careful with our terms. Take for example a solution made up by dissolving one formula weight of HA in water and making up the volume to one dm^3. What is the actual concentration of HA molecules? We cannot say that the solution is molar in HA in a strict sense because we know that a fraction α of the original formula weight of HA will have dissociated, hence the true concentration of HA is $(1 - \alpha)$ mol dm^{-3}. Thus we can see that in a solution made up of c formula weights of the acid per dm^3, the concentrations of both H^+ and A^- will be $\alpha \times c$. Thus the true concentration of undissociated acid HA is only $(1 - \alpha) \times c$.

These can now be inserted into the expression for K_a.

$$K_a = \frac{\alpha^2 c^2}{(1 - \alpha)c} = \frac{\alpha^2 c}{(1 - \alpha)}$$

You will recall that we are particularly interested in compounds which are not strongly dissociated, ie we expect α to be quite small.The simplifying assumption which we have already met is not uncommon in the use of equilibrium concepts by analysts, ie we assume that when α is much smaller than unity, the term $(1 - \alpha)$ approximates to 1.

The equation can of course be solved for α without this assumption. This involves rearranging it so that it is recognisable as a quadratic equation. You will note that the negative roots which occur in a purely mathematical treatment are meaningless. We will illustrate the problem by working through both methods.

(*a*) The first method utilises the assumption. It is another way of looking at the sort of problem posed in the ITQ above.

$$K_a = \frac{\alpha^2 c}{(1 - \alpha)}$$

If $\alpha \gg 1$ then $(1 - \alpha) \approx 1$

$$\therefore \quad K_a = \alpha^2 c \text{ and hence } \alpha = (K_a/c)^{\frac{1}{2}}$$

As an example we will take a solution of ethanoic acid containing 0.100 formula weights per dm^3; K_a for ethanoic acid is 1.82×10^{-5}.

$$\alpha = (K_a/c)^{\frac{1}{2}} = (1.82 \times 10^{-5}/10^{-1})^{\frac{1}{2}}$$

$$= 0.0135$$

(Note that α is dimensionless)

Π Calculate the pH of $0.100M$-ethanoic acid given that α is 0.0135.

$$[H^+] = \alpha c = 0.0135 \times 0.100M$$

$$= 0.00135M$$

Hence $\quad$ pH = 2.87

(*b*) Without the approximation.

Rearrange the expression thus:

$$K_a(1 - \alpha) = \alpha^2 c$$

ie

$$\alpha^2 c + K_a\alpha - K_a = 0$$

Recall that this is a quadratic equation of the type $ax^2 + bx + d = 0$, and use the standard formula for x.

$$a = c, \quad b = K_a, \quad d = -K_a$$

$$\alpha = \frac{-K_a + (K_a^2 + 4cK_a)^{\frac{1}{2}}}{2c}$$

$$= \frac{-1.82 \times 10^{-5} + (3.312 \times 10^{-10} + 0.728 \times 10^{-5})^{\frac{1}{2}}}{0.200}$$

Neglecting the root with negative sign

$$\alpha = \frac{-1.82 \times 10^{-5} + 2.698 \times 10^{-3}}{0.200}$$

$$\alpha = 0.0134$$

$$[H^+] = \alpha c = 0.00134\,M \quad ; \quad pH = 2.87.$$

Compare this with the value obtained by the approximation method.

To check that you understand this technique, work through the next SAQ with a view to answering these questions: is the simplifying assumption more or less valid,

(*a*) as K_a gets larger,

(*b*) as the concentration, c, gets smaller?

SAQ 2.7a

Calculate the percentage error introduced by the simplifying assumption in the calculation of $[H^+]$ for the solutions below.

(*i*) $K_a = 10^{-4}$, $c = 0.010M$

(*ii*) $K_a = 10^{-4}$, $c = 0.001M$

(*iii*) $K_a = 10^{-2}$, $c = 0.010M$

(*iv*) $K_a = 10^{-2}$, $c = 0.001M$

SAQ 2.7a

One practical difficulty occurs when referring to the concentrations of solutions 'as they are made up'. You will find that many texts approach this by saying – 'a solution containing so many grams per litre'. Another approach to this which is popular in some texts from the USA is the use of the term *formal concentration.* The formal concentration is the number of formula weights per dm^3; it takes the symbol F. It is therefore effectively the concentration in moles per dm^3 assuming no dissociation.

We can now return to our original problem and enquire about the effect of additional ethanoate ion on the ethanoic acid equilibrium. We shall use the symbol a for the molar concentration of added ethanoate and also assume that sodium ethanoate is completely dissociated.

As before we start with the expression for K_a and replace $[H^+]$ by αc and [HA] by $(1 - \alpha)c$ but this time $[A^-]$ is replaced by $\alpha c + a$. We make a similar simplifying assumption to that previously made and replace $\alpha c + a$ by a, ie

$$\alpha c \ll a$$

Hence
$$K_a = \frac{[H^+][A^-]}{[HA]} = \frac{\alpha c a}{(1 - \alpha)c}$$

$$= \frac{\alpha a}{(1 - \alpha)}$$

and if $\alpha \ll 1$, then the denominator reduces to 1 and $K_a = \alpha a$

or $\alpha = K_a/a$

Recall that $[H^+] = \alpha c$

Hence $[H^+] = cK_a/a$

For ethanoic acid, $K_a = 1.82 \times 10^{-5}$. We now look at the effect of added ethanoate ion at the following concentrations, $0.010M$, $0.100M$, and $1.00M$, with [ethanoic acid], $c = 1.0 \times 10^{-1}M$.

$$a = 0.010;\ [H^+] = \frac{1.82 \times 10^{-5} \times 1.0 \times 10^{-1}}{1.0 \times 10^{-2}} = 1.82 \times 10^{-4}M$$

$$a = 0.100;\ [H^+] = \frac{1.82 \times 10^{-5} \times 1.0 \times 10^{-1}}{1.0 \times 1{-}^{-1}} = 1.82 \times 10^{-5}M$$

$$a = 1.000;\ [H^+] = \frac{1.82 \times 10^{-5} \times 1.0 \times 10^{-1}}{1.00} = 1.82 \times 10^{-6}M$$

The pronounced effect of the common ion on the hydrogen-ion concentration is clearly seen. Addition of the common ion has depressed the hydrogen ion concentration. Check for yourself that Le Chatelier's qualitative approach gives the same general prediction.

Π Referring back to the question involving formic acid , what is the effect on the pH of 0.500M-formic acid when it is made 1.00M with respect to sodium formate? For formic acid $K_a = 1.77 \times 10^{-4}$. Specify any assumptions you make.

The assumptions are (*a*) that sodium formate is completely dissociated, hence the added formate ion concentration is the same as the formal concentration of sodium formate, (*b*) that the [formate ion] produced by dissociation of formic acid is negligible compared to [added formate].

You have already calculated the pH of formic acid alone, this was 2.03. Qualitatively we expect added formate to displace the dissociation equilibrium to the left-hand side, ie lower the hydrogen-ion concentration or increase the pH.

$$\text{Now} \quad [H^+] = \frac{K_a \times [HCOOH]}{[HCOO^-]}$$

$$[HCOOH] = 0.500M \text{ and } [HCOO^-] = 1.00M$$

$$\text{thus} \quad [H^+] = \frac{1.77 \times 10^{-4} \times 0.500}{1.00} = 8.85 \times 10^{-5}$$

$$\therefore \quad pH = 4.05$$

SAQ 2.7b

For the dissociation of HCN, $K_a = 4.93 \times 10^{-10}$. Calculate the pH of 0.10M HCN in water. What is the effect of making the solution (*a*) 1.00M in sodium cyanide, (*b*) 1.00M in HCl?

SAQ 2.7b

2.8 WEAK BASES AND THEIR SALTS

Reasoning analogous to that for weak acids and their salts may be used for equilibria involving weak bases and their salts. As in the previous example we shall look first at the free base, then consider the effect of added electrolytes. Aqueous ammonia is the most commonly met weak base and we now consider the effect of added ammonium chloride on this equilibrium system.

Recall that for this base $K_b = \dfrac{[NH_4^+][OH^-]}{[NH_3]}$

For ammonia there is some dispute about the actual nature of the undissociated species; ie whether it is more correctly formulated as $NH_{3(aq)}$ or NH_4OH. For our purposes the magnitude of K_b does not depend on the outcome of this debate. We might also note that the degree of dissociation is relatively small except at low concentrations.

By previous reasoning

$$[NH_4^+] = [OH^-] = \alpha c$$

and $$[NH_3] = (1 - \alpha)c$$

Hence $$K_b = \frac{\alpha^2 c^2}{(1 - \alpha)c} = \frac{\alpha^2 c}{(1 - \alpha)}$$

Eliminating α and rearranging,

Since $$\alpha = (K_b/c)^{\frac{1}{2}}, \text{ (assuming } \alpha \ll 1)$$

then $$[OH^-] = \alpha c = c(K_b/c)^{\frac{1}{2}}$$

or $$[OH^-] = (cK_b)^{\frac{1}{2}}$$

Π Use equilibrium concepts to calculat the pH of 0.100*M* aqueous pyridine. You are given pK_a for pyridine is 5.25.

Our first task is to convert the data supplied to a value for K_b.

Note however that the information given is in fact pK_a. This is not unusual in standard reference books, for example the data for weak organic bases in the *Handbook of Chemistry and Physics* (CRC Press) are all given in terms of K_a and pK_a, despite the fact that we are dealing with bases.

Recall that $$K_w = K_a \,.\, K_b$$

hence $$pK_w = pK_a + pK_b$$

As $$pK_w = 14, \; pK_b = 8.75 \text{ for pyridine,}$$

hence $$K_b = 1.78 \times 10^{-9} \text{ (antilog } - pK_b).$$

Now $[OH^-] = (cK_b)^{\frac{1}{2}}$

$= (1.78 \times 10^{-10})^{\frac{1}{2}}$

$= 1.33 \times 10^{-5}$

then $H^+ = 1 \times 10^{-14}/[OH^-] = 7.52 \times 10^{-10}$

hence $pH = -\log[H^+] = 9.12$

2.9. BUFFER ACTION

There are many situations during the work-up or preconcentration stages and during the analytical determination itself, when one wishes to have a controlled and essentially non-varying pH of one's solutions. However, we have seen that the pH is sensitive to such things as the presence of other compounds and dilution. We achieve the desired constancy by the use of buffer solutions. Buffer solutions are mixtures of compounds which in combination resist small changes in the hydrogen-ion concentration.The working solution is said to be buffered when a buffer solution has been added to it; the composition of the buffer is assumed to be *otherwise chemically innocuous*.

Many buffer mixtures are based on the mixed weak-acid and salt or weak base and salt concept. We sometimes have essentially the 'acid part' and the 'salt part' in the same molecule by working with partly neutralised polybasic acids. Buffers are available across a wide range of pH values and most analytical texts will contain data on their composition. It is important to remember that the phenomenon is based on the behaviour of equilibria and consequently it is temperature dependent.The change in pH over the range from zero to 50 ° C will often be less than one tenth of a pH unit, but sometimes it is greater. Borax solution for example, varies by *ca* 0.3 of a pH unit.

Some examples are given below. Note that concentrations are given both in molalities and in molarities.

	Molality	Molarity	pH
Potassium tetra-oxalate $KH_3(C_2O_4)_2.2\,H_2O$	0.0500	0.0496	1.68
Potassium tartrate (saturated at 25 °C) $KHC_4H_4O_6$	0.0340	0.0340	3.56
Potassium hydrogen phthalate $KHC_8O_4H_4$	0.0500	0.0495	4.01
Phosphate buffer *a*			
KH_2PO_4	0.0250	0.0249	
Na_2HPO_4	0.0250	0.0248	6.87
Phosphate buffer *b*			
KH_2PO_4	0.0087	0.00866	
Na_2HPO_4	0.0304	0.0302	7.41
Borax $Na_2B_4O_7.10\,H_2O$	0.0100	0.0099	9.18
Calcium hydroxide (saturated at 25 °C) $Ca(OH)_2$	0.0203	0.02025	12.45

Note: molarity is the number of moles per litre (dm^3) of solution, molality is the number of moles per 1000 g of solvent.

Fig. 2.2a. *Selected buffer solutions: pH values at 25 °C*

An important feature to remember about buffers is that the buffering power of solutions is limited to fairly small ranges of concentration of added acid or base. We refer to this feature as the buffer capacity. In general a solution will have a higher buffer capacity at higher concentrations of the buffering species. This will be developed quantitatively after an examination of the buffer effect itself as an equilibrium phenomenon. In addition it is worth noting that recent developments in electronics permit the measurement of pH with very great precision and that buffer systems do show minute changes of pH even within their normal range of application.

2.10. HYDROLYSIS OF SALTS

The introductory treatment of acids and bases usually represents the reaction of one with the other as a variation of the statement 'acid plus base gives salt plus water'.

Π Express this description of the reaction of an acid with a base as a generalised equilibrium. Recall that equilibria are reversible and write out the equilibrium for the reverse reaction.

What we end up with of course is simply:

acid + base $\rightleftharpoons$ salt + water, and

salt + water $\rightleftharpoons$ base + acid.

Be careful to note that we have said nothing specific about the acid or the base such as the K_a or K_b value. Therefore we cannot say anything about the position of the equilibrium. We can however say something about the position of the equilibrium if we are told that the acid and base are both strongly dissociated. Our equilibrium model tells us that if this is so, then the first reaction lies strongly to the right-hand side,and the second lies strongly to the left-hand side. We now wish to examine the situation represented by the second equation when the acid and/or the base is not strongly dissociated. This situation is commonly referred to as the *hydrolysis* of the salt.

It is most informative to examine the behaviour of the salt in terms of the ions which are formed on dissolution. For a generalised salt MX, ionisation will produce the ions M^+ and X^-. Dissociation of solvent water molecules will also ensure the presence of some H^+ and OH^- ions.

$$MX \rightleftharpoons M^+ + X^-$$

$$H_2O \rightleftharpoons H^+ + OH^-$$

You will see that the pair of ions H^+ and X^- could participate in an equilibrium which is identical to that for the dissociation of a weak acid.

$$HX \rightleftharpoons H^+ + X^-$$

Π What would be the equivalent conclusion to be drawn from considering the remaining ions?

A weak base may be formed.

We shall pursue this reasoning a little further by looking at an example for which the acid species HX has a small value of K_a, (check that you still remember what this means in terms of dissociation). Dissolution of the salt and dissociation of the solvent water have produced the following ions: M^+, H^+, X^-, and OH^-. However, if K_a is small then the predominant species will be undissociated HX rather than the ions. This is essentially the removal of H^+ ions from the system, ie a lowering of $[H^+]$. However we know that at constant temperature, the ionic product for water, $K_w = [H^+][OH^-]$ is constant, hence $[OH^-]$ must increase in response to this. Note that although we stated that the acid had a small value for K_a we did not say that K_b for the base was also small. Indeed if K_b is not small there will be no tendency to remove OH^- and the solution will be marginally alkaline because of this increase in the hydroxide-ion concentration.

We can now make a general statement about solutions of salts of a strong base and a weak acid: such solutions will have alkaline reactions. This process is frequently called the hydrolysis of the salt, although you will now see that it follows quite naturally from equilibrium considerations. Sodium ethanoate (sodium acetate) is a typical example of a salt of this type.

Π Go through a course of reasoning similar to that above for the salt of a strong acid and a base with a small K_b value.

Here we form weakly dissociated MOH, consequently lowering the hydroxide-ion concentration. Because of the constancy of K_w there is an increase in hydrogen-ion concentration, ie salts of strong acids and weak bases have an acidic reaction. Again this is described as hydrolysis.

We complete our examination of the system by considering salts formed from strong acids and strong bases and salts formed from weak acids and weak bases. When both the acid and the base are strongly dissociated there is an *equal* and very small drive to form undissociated species. When both the acid and base are weakly dissociated there will be a *similar* and high drive to form undissociated species. In both cases the *ratio* of $[H^+]$ to $[OH^-]$ remains unchanged, and thus the solutions remain neutral.

Salt formed from: acid	base	Dissociation constants	Reaction of aqueous solution	Examples
strong	strong	$K_a \simeq K_b$	neutral	KNO_3;NaCl
strong	weak	$K_a \gg K_b$	acidic	NH_4Cl;pyHCl
weak	strong	$K_a \ll K_b$	alkaline	NaAc;NaBz
weak	weak	$K_a \simeq K_b$	neutral	NH_4Ac;NH_4Fm

py = pyridine, Ac = ethanoate,
Bz = benzoate, Fm = Formate (methanoate).

This approach allows us to make a general classification in terms of individual ions. For example anions derived from weak acids are basic in character and cations derived from weak bases are acidic. Anions derived from strong acids and cations derived from strong bases are neutral. Polybasic acids will also give rise to acidic anions because of their partial ionisation. These are included in the table below for completeness.

	Acidic	Neutral	Basic
Anions	HSO_4^-	Cl^-	CN^-
	$H_2PO_4^-$	NO_3^-	F^-
		Br^-	HS^-
		ClO_4^-	CO_3^{2-}
		I^-	NO_2^-
		SO_4^{2-}	PO_4^{3-}

	Acidic	Neutral	Basic
Cations	Mg^{2+}	Li^+	None
	Al^{3+}	Na^+	
	NH_4^+	Ca^{2+}	
	Zn^{2+}	K^+	
		Ba^{2+}	

SAQ 2.10a

Classify the following salts as 'acid', 'basic' or 'neutral' according to their behaviour in water:

KCN, $NaNO_3$, NaH_2PO_4, $Zn(ClO_4)_2$

Give the relevant equations when they are not neutral.

The Hydrolysis Constant

The ideas described above can be made more quantitative by treating the process of hydrolysis in a manner analogous to the treatment of the equilibria which gave rise to K_a, K_b and K_w. This means introducing a *hydrolysis constant*, K_h. For a salt of a strong acid and a weak base we have the equilibrium below; note that the anion would appear on both sides of the equation and is therefore omitted.

$$B^+ + H_2O \rightleftharpoons BOH + H^+$$

$$K_{eq} = \frac{[BOH][H^+]}{[B^+][H_2O]}$$

As before we assume that at low concentrations activity coefficients approach unity. We also assume, as before, that the concentration of water is constant and that it is possible to replace $K_{eq} \times [H_2O]$ by a new constant. This becomes our definition of K_h.

$$\therefore \qquad K_h = \frac{[BOH][H^+]}{[B^+]}$$

We now relate this to the terms K_a and K_b, with which we are already familiar, by means of a little algebra.

Both the numerator and denominator are multiplied by $[OH^-]$

$$K_h = \frac{[BOH][H^+][OH^-]}{[B^+][OH^-]}$$

You will recall that $K_w = [H^+][OH^-]$ and that $K_b = [B^+][OH^-]/[BOH]$, hence we have:

$$K_h = \frac{K_w}{K_b}$$

SAQ 2.10b Use reasoning similar to that above to derive an expression relating K_h for a salt of a strong base and weak acid to K_w and K_a.

SAQ 2.10b

These expressions for K_h allow the hydrogen-ion ion concentration of solutions to be computed over a range of concentrations so long as K_b (or K_a) and K_w are known at the temperature of the experiment.

For concentration c, we start by recalling that [BOH] and $[H^+]$ are equal as required by the stoichiometry. Also to a first approximation we can replace $[B^+]$ by c; (ie we neglect the small lowering of $[B^+]$ which arises because of the formation of BOH, strictly $[B^+] = c\text{-}[BOH]$).

$$\text{Thus} \quad K_h = \frac{K_w}{K_b} = \frac{[BOH][H^+]}{[B^+]} = \frac{[H^+]^2}{[B^+]} = \frac{[H^+]^2}{c}$$

$$\text{ie} \quad [H^+]^2 = cK_w/K_b$$

$$\text{or} \quad [H^+] = (cK_w/K_b)^{\frac{1}{2}}.$$

If you now take logs you will see that:

$$pH = \tfrac{1}{2}pK_w - \tfrac{1}{2}pK_b - \tfrac{1}{2}\log c.$$

SAQ 2.10c

By using reasoning similar to that above, show that in the hydrolysis of a salt of a weak acid and a strong base, pH is given by:

$$\tfrac{1}{2}pK_w + \tfrac{1}{2}pK_a + \tfrac{1}{2}\log c.$$

We must remember that the phenomenon of hydrolysis is concentration dependent. We can demonstrate this by using a term 'the degree of hydrolysis', x, and considering the behaviour of 1 mole of BX in a volume v dm^3 (x is the fraction of the salt hydrolysed).

$$K_h = \frac{[BOH][H^+]}{[B^+]}$$

$$= \frac{(x/v)(x/v)}{(1 - x)/v}$$

Hence

$$K_h = \frac{x^2}{v(1 - x)}$$

Note that here we are using a more correct form for $[B^+]$, ie as $c = 1/v$, then $[B^+] = (1 - x)/v$.

You will see that as v increases, ie the solution becomes more dilute, then x also increases, as K_h is a constant. That is, one predicts more hydrolysis in the more dilute solutions; conversely one would work in more concentrated solutions in order to depress hydrolysis.

Objectives

Now that you have completed this Part you will now be able to:

- identify autoprotolysis,
- describe acids and bases in Brönsted–Lowry terms,
- identify conjugate acids and bases,
- recognise neutralisation in terms of conjugate acid–base pairs,
- use K_w to calculate pH,
- formulate a relationship between K_w, K_a and K_b,

- calculate pH at various stages of a neutralisation titration,
- calculate the effect of common ions on pH,
- understand simplifying assumptions in the treatment of weak acids and to calculate the likely errors associated with this,
- describe buffer action in equilibrium terms,
- obtain the hydrolysis constant in terms of K_w and K_a or K_b.

3. Solubility

This Part describes the concept of solubility as it is used by the analyst and relates this to the equilibrium model and the concept of the solubility product. The practical nature of the analyst's interest in solubility implies the usual presence of common ions, and the common-ion effect as developed *via* the equilibrium model.

3.1. EQUILIBRIUM LAW

Much of analytical chemistry is concerned with the behaviour of solutions and the solubility of solutes in solvents. These considerations may be relevant to the collection of samples, the concentration of samples and the actual analytical measurements themselves. Further we may be interested in the processes by which materials dissolve and those which lead to precipitation. The equilibrium model has proved to be very useful in describing these and an introductory treatment of these phenomena is given below. An amplification of these ideas will be found in Parts 8 and 9 on precipitation methods of analysis.

If we consider a slightly soluble ionic solid, MX, in the presence of water as solvent, the solid enters the solution phase as $M^+(aq)$ and $X^-(aq)$. The symbol (aq) is used to indicate that the ions are hydrated (solvated), but it is not intended to indicate any particular number of water molecules per ion. We shall omit these subscripts in the treatment below for simplicity, but do not forget that solvent species are always present. When the solid is added to the solvent water, the system is not at equilibrium as M^+ and X^- dissolve. Eventually no more ions are dissolved and, assuming that MX is present in excess, the solution is saturated. Solid MX is now in equilibrium with dissolved M^+ and X^-.

$$MX_{(solid)} \rightleftharpoons M^+ + X^-$$

We can now write an equilibrium expression:

$$K_{eq} = \frac{[M^+][X^-]}{[MX]}$$

But MX is present as a solid and you will recall that we can assume its activity to be constant irrespective of the amount of solid present. This remains true as long as there is some solid MX present in equilibrium with the solution.

Thus $$K_{eq} = \frac{[M^+][X^-]}{\text{constant}},$$

assuming as usual, that activities and concentrations are equal.

We can use the technique which we have met before in which two constant terms are combined and a new term defined. Here we redefine (K_{eq} × constant) as K_{sp}, the *solubility product constant*. This is frequently called simply the *solubility product*. (You may also find that some texts use the symbol S_{MX} instead of K_{sp}. Be careful not to confuse this with the solubility of MX which is also written S_{MX})

ie $$K_{sp} = [M^+][X^-]$$

Look closely at the terms which make up K_{sp} and you will see that the units here are M^2 or $mol^2\ dm^{-6}$. You must also be careful to remember that the values of K_{sp} are constant only at constant temperature. This follows if you recall that K_{eq} will also change with temperature.

Π Derive similar but not identical expressions for K_{sp} for the following species and give their units in the convention using M^2 rather than $mol^2\ dm^{-6}$ etc; M_2X, MX_2 and M_3X.

The ionisation equations are:

$$M_2X \rightleftarrows 2\,M^+ + X^{2-}$$

$$MX_2 \rightleftarrows M^{2+} + 2\,X^-$$

$$M_3X \rightleftarrows 3\,M^+ + X^{3-}$$

Consequently the K_{sp} expressions are:

M_2X; $K_{sp} = [M^+]^2[X^{2-}]$; M^3

MX_2; $K_{sp} = [M^{2+}][X^-]^2$; M^3

M_3X; $K_{sp} = [M^+]^3[X^{3-}]$; M^4

3.2. THE SOLUBILITY PRODUCT

You will recall that we frequently express solubility in terms of the molar concentration of the solute in the solvent. Consequently we can confidently expect to be able to derive a relationship between the solubility S and K_{sp}.

For compounds of the type MX, S the molar concentration of MX in solution as M^+ and X^-, must be the same as the concentration of M^+, which here is the same as that of X^-.

Thus $$K_{sp} = [M^+][X^-] = [M^+]^2 = [X^-]^2 \text{ and}$$

$$[M^+] = K_{sp}^{\frac{1}{2}}$$

ie $$S = K_{sp}^{\frac{1}{2}}$$

Note that this leaves the units correct, as for MX, K_{sp} had units of M^2, and we know S has units of M.

SAQ 3.2a

> **Given that the solubility product of copper(I) chloride at 25 °C is $1.20 \times 10^{-6} M^2$, calculate the solubility of copper(I) chloride at equilibrium in water at 25 °C.**
>
> **If you had a sample which originally contained 1.0 g CuCl in contact with one dm^3 of water at 25 °C, what would be the approximate percentage error introduced by assuming that copper(I) chloride was completely insoluble?**

The loss to solution by the slight solubility of a substance commonly regarded as insoluble noted in the SAQ above seems surprisingly high. However in practice there are processes, as we shall see below, which allow this to be very much reduced.

Before passing on to this we shall look briefly at the related solubility calculations for other types of compound, eg the solubility of compounds of the type MX_2

For MX_2 $K_{sp} = [M^{2+}][X^-]^2$

Now when one mole of MX_2 dissolves we get two X^- ions for each M^{2+} ion,

$$\therefore \quad [X^-] = 2[M^{2+}]$$

and

$$S = [M^{2+}]$$

We can now express K_{sp} in terms of $[M^{2+}]$,

$$K_{sp} = [M^{2+}] \times 2[M^{2+}] \times 2[M^{2+}]$$

$$= 4[M^{2+}]^3$$

$$\therefore \quad S = [M^{2+}] = (K_{sp}/4)^{\frac{1}{3}}$$

SAQ 3.2b

> **Use a treatment similar to that above to obtain an expression for the solubility of a compound MX_3 in terms of K_{sp}.**

SAQ 3.2b

3.3. THE COMMON-ION EFFECT AND SOLUBILITY

We have already met the common-ion effect in the context of the dissociation of weak electrolytes. Here we shall examine the effect in relation to the solubility of sparingly soluble substances.

We can predict the qualitative nature of the common ion effect by recalling the equilibrium in a saturated solution in contact with undissolved solute.

$$MX_{(s)} \rightleftharpoons M^+_{(sol)} + X^-_{(sol)} \quad [(sol) = \text{in solution}]$$

Now Le Chatelier's principle predicts that an increase in the concentration of one of the ions which is present in MX, say X^-, will cause the equilibrium to move to the left-hand side of the equation. That is an externally added common ion *reduces* the solubility of the compound. This will be true whether the ion added is X^- or M^+. We can also use the same reasoning to predict an increase in solubility in situations which would lower the concentration of one of the ions present. These ideas are made quantitative simply by including known values for the concentrations. Note however that as before we make a simplifying assumption, *viz* in the presence of large amounts of added species the contribution from the dissolution of the low-solubility species is negligible.

SAQ 3.3a

Given that $K_{sp}(AgCl) = 1.80 \times 10^{-10}$ mol^2 dm^{-3} at 25 °C, calculate the solubility of silver chloride in the following solutions:

(*i*) pure water,
(*ii*) 0.0010*M*-KCl,
(*iii*) 0.0100*M*-KCl,
(*iv*) 0.1000*M*-KCl.

SAQ 3.3b

What is the molar concentration of dissolved barium chromate in a solution which is made 0.060 *M* in barium chloride? ($K_{sp} = 1.2 \times 10^{-10} M^2$).

SAQ 3.3b

In the examples above we can make a simplifying assumption about the concentration of one of the species. When it is not possible to make this assumption calculations can still be carried out but they become a little more involved. The following exercise illustrates this.

Π What is the concentration of thallium $[Tl^+]$ present in water at equilibrium with solid thallium iodide in 10^{-4} M sodium iodide. $K_{sp}(TlI) = 6.00 \times 10^{-8} M^2$?

In this case total $[I^-]$ = [added I^-] + [I^- from TlI].

However the iodide from thallium iodide has the same concentration as the thallium ion itself.

$$K_{sp} = [Tl^+][I^- \text{ total}]$$

$$= [Tl^+][10^{-4} + Tl^+]$$

$$= 6.00 \times 10^{-8}$$

This expands to,

$$[Tl^+]^2 + [Tl^+] \times 10^{-4} - 6.00 \times 10^{-8} = 0$$

We can use the standard quadratic equation formula to solve this for $[Tl^+]$.

Hence $[Tl^+] = 2.00 \times 10^{-4} M$.

3.4. IONS WHICH ARE NOT COMMON WITH THOSE OF THE MAIN SOLUTE

As we have seen that the addition of a common ion decreases the solubility of precipitates, it is a reasonable extension of the equilibrium model to ask whether it has anything to say on the presence of electrolytes which do not contain a common ion. To do this we must go back to one of our early simplifying assumptions; ie that activities are adequately approximated to by concentrations. We have seen that in an analytical context this is an acceptable assumption if concentrations are low.

However to express K_{sp} as a true constant we should use activities.

$$K_{sp} = a_m \times a_x \text{ where } a = \text{activity.}$$

ie $K_{sp} = [M^+]f_m \times [X^-]f_x$, where f = activity coefficient.

Rearrangement gives

$$[M^+][X^-] = K_{sp}/(f_m f_x)$$

When an electrolyte is added to the solution its ions are solvated and thus some solvent molecules are effectively 'removed' from the system. This has the effect of *decreasing* f_m and f_x as the concentration of the added electrolyte increases. Thus we can see that decreasing the activity coefficients will *increase* the concentration of $[M^+]$ and of $[X^-]$. Therefore we can see that the equilibrium law predicts that the presence of high concentrations of electrolytes without a common ion will increase the solubility of sparingly soluble salts.

These ideas are developed more fully in the Parts on precipitation methods of analysis.

Objectives

Now that you have completed this Part you will be able to:

- relate K_{sp} to K_{eq},
- derive expressions for K_{sp} for compounds of several formulae,
- relate K_{sp} and solubility, S,
- carry out calculations relating S and concentrations of common ions.

4. Complex Ion Formation

This Part gives a brief description of the process of complex ion formation and its effect on metal ions in aqueous solutions. The equilibrium model is applied to this process and the concept of the formation constant developed. This is then extended to include stepwise processes. Titration curves for complexometric titrations are derived and the importance of K values is developed. Chelation is described and the scope for application of EDTA is described *via* the equilibrium concept.

4.1. COMPLEX IONS

Before examining the significance of the equilibrium concept for the formation of complex ions, it is as well to clarify just what we mean by a *complex ion.* As a start we must be careful not to confuse 'complex' with 'complicated', even though some complex species are in fact of complicated structure. Also, although the term 'complex' is very common in chemical parlance, both as a noun and a verb, it is instructive if we start with the word 'coordination' which has essentially the same meaning. You will see that we may refer to a complex ion or complex as a coordination compound. The process of complex formation or complexation is identical with that of coordination of one or more donor groups with a metal ion.

The process of coordination is best illustrated by a well known example. You will be familiar with the reaction of copper(II) sulphate solution with an *excess* of aqueous ammonia. Addition of ammonia to the pale-blue solution leads initially to a light-blue precipitate of copper(II) hydroxide, which redissolves to give a deep-blue solution. We say that the ammonia molecules have coordinated with the copper ion to give the complex ion $Cu(NH_3)_4^{2+}$, ie the cation is formed by combination of a central metal ion with other groups.

In bonding terms, this occurs because the copper(II) ion has vacant orbitals by means of which it can accept electrons from donor groups.Thus the metal ion is potentially a weak acid in the Lewis classification. The ammonia molecule has a non-bonded or lone pair of electrons and is a weak base as we have already seen: hence it can act as a donor to the metal. It is therefore acting as a Lewis base, or using the terminology of organic chemists, it is a nucleophile.

In the vocabulary of coordination chemistry we call the donor ions or molecules *ligands*. Thus the process loosely called coordination or complexation is in fact the coordination of donor species or ligands to the metal ion. The product is called *the complex* and may have either a positive charge, a negative charge, or have no charge depending on the charge of the free ligand. Examples follow.

$$Cu^{2+} + 4\,NH_3 \rightleftharpoons Cu(NH_3)_4^{2+}$$

$$Cu^{2+} + 4\,CN^- \rightleftharpoons Cu(CN)_4^{2-}$$

$$Cu^{2+} + 2\,NH_2CH_2COO^- \rightleftharpoons Cu(NH_2CH_2COO)_2$$

In these examples the compound isolated is called a *coordination compound*, but in practice chemists also loosely call these *complexes*; here, these are $Cu(NH_3)_4SO_4$, $K_2Cu(CN)_4$, and $Cu(NH_2CH_2COO)_2$.

You will note that we have represented these reactions as equilibrium processes, but it should not have escaped your notice that there is more than one ligand per metal ion. We should be asking questions such as:

'Do the ligands all go on at once or is it a stepwise process?'

'What other sorts of ligands will do this?'

'Do the properties exhibited by the metal ion change on complexation?'

Because coordination chemistry is a vast corpus of knowledge we intend here to draw out only those features which are relevant to analysis, and to relate these to the equilibrium ideas which were developed at the beginning of this section.

We shall take the questions in reverse order and illustrate their significance for various aspects of analytical chemistry.

SAQ 4.1a

Decide whether each of the following statements is true or false.

(*i*) When a coordination compound is formed the ligand is behaving as a Lewis acid

T / F

(*ii*) The equilibria involved in the formation of complex species usually involve the displacement of one donor group by another.

T / F

(*iii*) Ligand molecules always contain neutral donor groups.

T / F

SAQ 4.1a

SAQ 4.1b

The addition of ammonia to $CoCl_3$ gives rise to the compound below. Identify the components by marking the grid as appropriate.

$Co(NH_3)_6Cl_3$ or

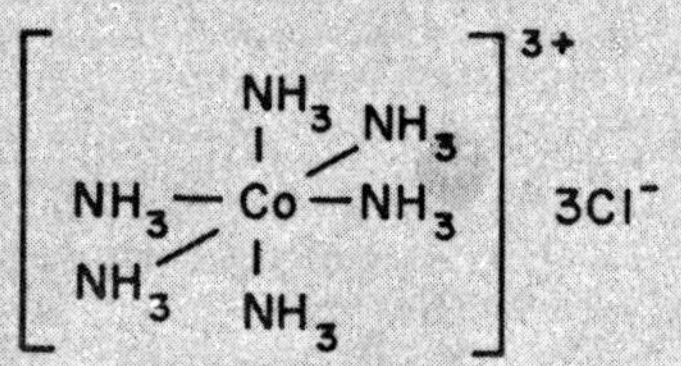

Cl^- NH_3 Co^{3+} $Co(NH_3)_6^{3+}$ $Co(NH_3)_6Cl_3$

ligand,

complex-ion,

nucleophile,

coordination
-compound,

Lewis acid,

Lewis base.

SAQ 4.1b

4.1.1. The Properties of the Metal Ion

You will recall that in our example we had the insoluble copper(II)hydroxide apparently rendered soluble by the addition of the ammonia and that the intensity of the colour was vastly increased and the colour itself changed. Another familiar example is the increase in solubility of silver chloride brought about by the presence of sodium iodide, in this case due to the formation of the ion, $AgClI^-$.

A similar example is the failure of chloride ion to precipitate silver chloride in the presence of aqueous ammonia because of the formation of a silver ammine complex (note that complexes containing NH_3 are called *ammines* – double 'm' – in contrast to compounds of the type RNH_2 which are the familiar amines). You will appreciate the potential use in analytical chemistry of making compounds more soluble than they otherwise would be.

We might also note here that when the complex ion is charged it is fairly likely that it will be at least to some extent soluble in water. On the other hand, when the complex species is neutral it is expected that it will have a very low solubility in water but may well be quite soluble in non-aqueous solvents. Again one can see the potential for exploiting these features for extractive and separative techniques. For both a knowledge of equilibria will assist in understanding these phenomena.

Other properties which we expect will change after the complexation are, for example, conductivity (again related to the charge type) and the electrode potentials for species which undergo redox behaviour (discussed in Part 5). Yet another type of application can be guessed from our original example of the copper(II)-ammonia system, that is exploitation of the colour change. Some metal complexes have very intense absorption in the visible region of the spectrum and this has led to their application as metal-ion indicators in classical wet analysis, and as coloured species in instrumental analysis.

4.1.2. Other Sorts of Ligand

There is a multitude of types of ligand available for complexing metal ions.

The ligand molecule itself may have one or more potential coordinating positions, these are referred to by a group of terms reminiscent of 'teeth' – the analogy being the the number of 'biting' sites on the ligand. Thus there are unidentate ligands (one binding site) (also called monodentate), bidentate ligands (two binding sites), and poly-dentate ligands(several binding sites). Unidentate ligands can of course be neutral, eg NH_3, Ph_3P, H_2O; or charged, eg Cl^-, OH^-.

The bi-and poly-dentate ligands offer an enormous range of possibilities for ligand types,some with neutral coordinating groups, some with charged coordinating groups and some with both types of group. A few examples are listed in Fig. 4.1a: note that abbreviations and non-systematic names are given.You will find that these are in fact the names commonly used by analysts.

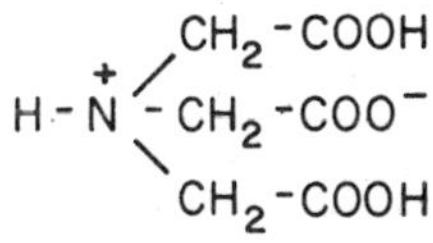

Nitrilotriacetic acid
(NITA or NTA)

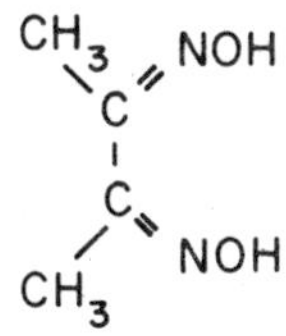

Butane-2,3-dione dioxime
(Dimethylglyoxime or DMG)

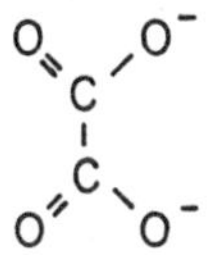

Oxalate ion
(Ethanedioate ion)

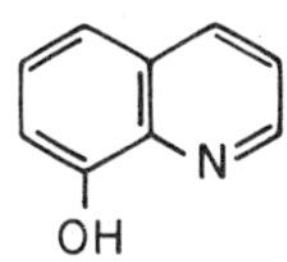

8-Hydroxyquinoline
(Oxine)

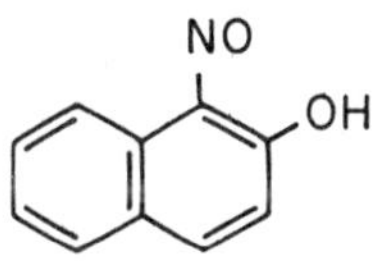

1-Nitroso-2-naphthol

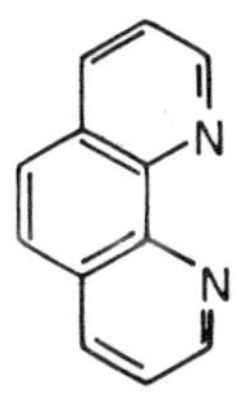

1,10-Phenanthroline
(*Ortho*-phen or *o*-phen)

Fig. 4.1a. *Some examples of ligands used in analytical chemistry*

In addition you will find that some ligands are commonly referred to as 'better' or 'stronger' than others, similarly described as 'poorer' or 'weaker'. These terms refer to the general stability of the complexes formed and to the ability of some ligands to displace others. A useful rule of thumb is that the ligands become increasingly poor in the following order:

$$CN^- > \text{bidentate ligands} > \text{N-donors} > H_2O \simeq \text{halides}$$

You will recognise that we can approach this phenomenon *via* our equilibrium model. The stability of the complexes formed will be related to the *equilibrium constant* for the formation reaction of the metal with the ligands, ie the 'better' ligands will give rise to high values for K_{eq}. You will find that the chemical literature and inorganic chemists refer to these as 'stability constants' or 'formation constants'. One important consequence of this hierarchy of ligand strengths is the possibility of progressive replacement of one type of ligand by another. In fact in aqueous solution we should strictly regard the metal ion as a complex in which the ligand is water. Thus the conventional process of complex formation is really one of displacement of a weak ligand, water, by a stronger ligand, eg NH_3.

However it is probably in the field of complex formation between metal ions and amino-polycarboxylic acids (poly-dentate ligands) that the process of complexation has had most impact on the practice of analytical chemistry. This is covered by separate Parts dealing with detailed and specific applications.

Our third question was 'Do the ligands go on at once or in a stepwise fashion?' In order to answer this we need to apply our equilibrium model to the system in a little more detail.

4.2. FORMATION CONSTANTS AND COMPLEXES

We start by examining a single step, the reaction of a metal ion M, with one ligand molecule, L. Charges and hydration water are omitted for simplicity.

Π How would you express the equilibrium constant for this?

The equation is:

$$M + L \rightleftarrows ML$$

therefore $$K_{eq} = \frac{[ML]}{[M][L]}$$

However if a second ligand molecule can be added we get a second equilibrium and of course a second equilibrium constant. In order to identify these properly we will call them K_1, K_2, etc.

Π Write out the expression for K_2 for the M/L system.

The reaction is:

$$ML + L \rightleftharpoons ML_2$$

therefore

$$K_2 = \frac{[ML_2]}{[ML][L]}$$

We can see that this process can be repeated until no more ligand molecules can be accommodated around the central metal ion, ie the complex of maximum coordination number has been reached.

In fact, the formation of ML_n occurs in *stepwise fashion.* The equilibrium constants K_1, K_2, ... K_n are called *stepwise formation constants* (or occasionally stepwise stability constants). In practice, we can often observe this stepwise formation, ie the individual species ML, ML_2, ... ML_n are observable. The familiar Cu^{2+}/NH_3 system is an example of this in which species up to $n = 5$ are observed in aqueous solutions. The value of K_6 is so small that liquid ammonia systems are needed to permit the observation of the species with $n = 6$. In other systems the adjacent K values are so close together that separate observation of all intermediate species is not possible. We shall return to this shortly.

Π Write out the expressions for all the K values up to a maximum coordination number of four and examine the result of multiplying all of these together.

If you examine the expressions for the K values you will see that many of the terms appear in both the numerators and the denominators of the complete set of expressions.

$$K_1 = \frac{[ML]}{[M][L]}$$

$$K_2 = \frac{[ML_2]}{[ML][L]}$$

$$K_3 = \frac{[ML_3]}{[ML_2][L]}$$

$$K_4 = \frac{[ML_4]}{[ML_3][L]}$$

So the product $K_1.K_2.K_3.K_4$ is given by:

$$\frac{[ML]}{[M][L]} \times \frac{[ML_2]}{[ML][L]} \times \frac{[ML_3]}{[ML_2][L]} \times \frac{[ML_4]}{[ML_3][L]}$$

On appropriate cancellation this reduces to:

$$\frac{[ML_4]}{[M][L]^4}$$

Let us now look at the expression we would get if the ligands all went on to the metal ion at once. The reaction is:

$$M + 4L = ML_4$$

hence $$K_{eq} = \frac{[ML_4]}{[M][L]^4}$$

We call this the *overall stability constant*, and assign it a separate symbol β (beta). In the example above this would be β_4.

It is worth noting here that the normal usage for the representation of complex formation is in fact the reverse of that used for acid-base equilibria. You will recall that we have discussed acids and bases in terms of dissociation equilibria:

$$HA \rightleftharpoons H^+ + A^-$$

An analogous treatment of a complex ion would be a dissociation of the following type:

$$ML \rightleftharpoons M + L$$

Thus some texts refer to *instability constants* (ie the equilibria are expressed this way round) rather than the more common terms *formation constant* or *stability constant*. The latter two terms are used interchangeably.

SAQ 4.2a

For the equilibrium involving Ni^{2+} and NH_3 write out the expressions for the following, omitting any participation by water molecules:

K_1, β_2, K_3, and β_6.

Because of the possibility of the formation of several complexes *via* the stepwise series of equilibria, the experimental determination of equilibrium constants is rather involved. When it is known that only one product species is present, the determination of the necessary concentrations of M, L and ML_n is relatively easy. This may be by spectroscopic or polarographic means, or if the ligand is a weak acid, by pH measurement. However there is usually more than one complex species present and a large number of measurements covering a wide range of concentration ratios is necessary for the evaluation of consecutive formation constants.

Consecutive formation constants for unidentate ligands usually decrease as the number of ligands increases. A consequence of this is that there will be a reduced tendency to add an extra ligand as the number of ligands already present increases, ie it is 'hardest to get the last one in'. As this has important implications for analytical applications we shall examine it in more detail.

If we take a mixture of a metal ion, M, and a ligand, L, in a molar ratio of say 1 : 3 we shall have a mixture of the species up to ML_n where n is the maximum coordination number. The exact proportions of these will depend on the relative magnitude of all the K values from K_1 to K_n. A typical example is the system of Cu(I) and cyanide ion (the coordination maximum is four). A mixture of Cu^+ and CN^- in the ratio 1 : 3 contains about 20% of $Cu(CN)_2^-$, 60% of $Cu(CN)_3^{2-}$, and 20% of $Cu(CN)_4^{3-}$. Only if the stepwise formation constants are widely different shall we expect dominance by one complex species alone. Of course we can see that very large excess concentrations of the ligand will also favour the formation of the complex with the highest coordination number, but even this is at times not sufficient to guarantee formation of the complex with the maximum possible value of n.

We can appreciate the significance of this feature if we recall that in many analytical methods we may be measuring a property of a complex species. Unless we are certain that the material we wish to determine is all in the same chemical form measurements are meaningless, as we would be gathering results on a non-specific mixture.

Therefore many prescribed methods of analysis call for moderate-to-large excesses of ligands to ensure that the analyte is all in the same coordinated form.

We now take a specific example, the complexes of nickel ions and ammonia.

The stepwise formation constants for $Ni(NH_3)_n$ are:

$$K_1 = 1 \times 10^{2.79} = 6.17 \times 10^2$$

$$K_2 = 1 \times 10^{2.26} = 1.82 \times 10^2$$

$$K_3 = 1 \times 10^{1.69} = 0.49 \times 10^2$$

$$K_4 = 1 \times 10^{1.25} = 0.18 \times 10^2$$

$$K_5 = 1 \times 10^{0.74} = 0.055 \times 10^2$$

$$K_6 = 1 \times 10^{0.03} = 0.011 \times 10^2$$

The close proximity of neighbouring values for K_1 to K_5 means that the exact nature of the mixture of species present when ammonia is added to a solution containing Ni^{2+} ions will depend on the ratio $[Ni^{2+}]/[NH_3]$. As we increase the amount of ammonia present we expect a smooth transition from a range of species of low values of n through to $n = 5$. However because of the larger jump in value between K_5 and K_6 it turns out that the principle component of mixtures with a large excess of ammonia is still $Ni(NH_3)_5^{2+}$.

We might also note here that our equilibrium model leads us to expect some dissociation *via* NH_4OH to give NH_4^+ and OH^-. In practice we can suppress this by working at high concentrations of added ammonium salts, again as explained by our equilibrium model.

The phenomenon of consecutive formation constants with relatively close values has another feature of interest to us. This is in the application of titration methods to the determination of metals by using complexing ligands, known as complexometric methods. These methods are dealt with in detail in Part 13.

If we consider the titration of a metal-ion solution by using a unidentate ligand as the titrant, we expect the titration curve to be as shown in Fig. 4.2a.

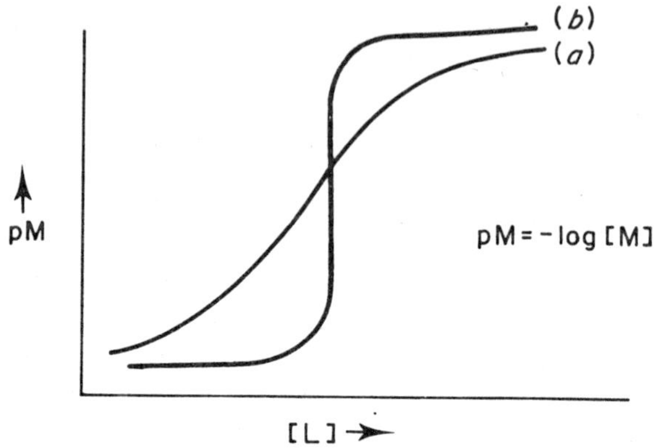

Fig. 4.2a. *Hypothetical titration curves for M/L systems forming ML_n, when (a) K decreases as n increases, and (b) K increases as n increases*

(*a*) This plot of $-\log[\text{M}]$ against added [ligand] will be unlikely to show a sharp end-point. This is because the added ligand can be distributed over the range of complexes possible. Thus complex formation of this sort is not likely to be a candidate for complexometric titrations, even with sophisticated end-point detection.

(*b*) If however, we had a sequence for which the later *K* values exceed the earlier *K* values, then the principal species formed would be the higher complex and we might realistically expect a conventionally shaped titration curve permitting sharp end-points. The Ag^+/NH_3 system has $K_2 > K_1$ and so might be a possible candidate for this approach. However there is a group of ligands which is by far better suited to this purpose. This group of ligands which we examine below has been of exceptional importance in analytical chemistry.

SAQ 4.2b

Which of the following is the correct concluding phrase to the phrase below?

'In analytical procedures using unidentate ligands it is common to require large excesses of the ligand because ... '

(*i*) ... large excesses of all reagents are good laboratory practice.'

(*ii*) ... one always needs large excesses to form any coordination compound.'

(*iii*) ... large excesses are frequently needed to ensure the predominance of only one complex species.'

(*iv*) ... stability constants are always small for coordination compounds.'

SAQ 4.2c

For the hypothetical series of complex equilibria involving stepwise reaction of a ligand L with a metal M in the molar concentration ratio L:M of 4:1, which of the following descriptive phrases would be expected to match the data?

	A	B	C	D
$K_1 =$	10^6	10^3	10^3	10^5
$K_2 =$	$10^{0.7}$	$10^{2.8}$	10^4	$10^{4.5}$
$K_3 =$	$10^{0.5}$	$10^{0.1}$	$10^{4.5}$	10^4
$K_4 =$	$10^{0.1}$	$10^{0.02}$	10^6	$10^{3.5}$

(*i*) There should be a mixture with significant concentrations of all species, ML, ML_2, ML_3, and ML_4 along with unbound L.

(*ii*) The species ML_4 will predominate.

(*iii*) The mixture will be largely ML and ML_2 with some unbound L.

(*iv*) There will be a significant amount of unbound L and the principal complex will be ML.

SAQ 4.2c

4.3. THE CHELATE EFFECT

The thermodynamic stability of complexes may be represented by the relationship between the standard free-energy change, ΔG° and β_n, below. (The superscript $^\circ$ indicates a standard change).

$$\Delta G^\circ = -RT \ln\beta_n$$

You may recall from previous studies that the free-energy change, ΔG, is represented by a combination of two other terms. One is called the enthalpy change, ΔH, in general terms the heat-change part, and the other is the entropy change, ΔS, the disorder-change part.

$$\Delta G^\circ = \Delta H^\circ - T\Delta S^\circ$$

where T is the temperature in Kelvins.

Although the enthalpy part is usually the more important contribution to the stability of complexes, sometimes the entropy part is dominant. We need to bear these two terms in mind when examining a phenomenon known as the *Chelate Effect*. The name of this effect comes from the Greek word for a crab's claw (Fig. 4.3a), and the analogy is that the ligand behaves like a crab, that is with two more or less opposed points of attack. Clearly ligands which display this effect must have at least two coordination centres, ie be bidentate or polydentate. We use the word 'chelation' in fact for any coordination higher than that of a unidentate ligand.

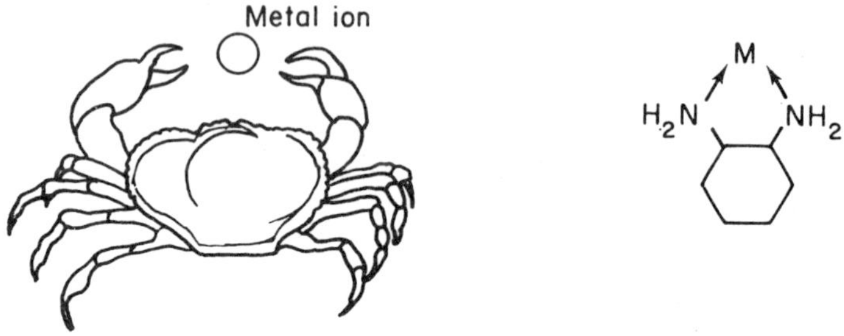

Fig. 4.3a. *Chelating nature of bidentate ligands, eg 1,2-diaminocyclohexane*

We can illustrate the effect by comparing the binding of a metal ion with, say, four unidentate ligands with that of two bidentate ligands of broadly similar structure. These are the unidentate ligand methylamine, CH_3NH_2, and the bidentate ligand 1,2-diaminoethane, $NH_2CH_2CH_2NH_2$, (often given the symbol 'en' on account of its non-systematic name, ethylenediamine)

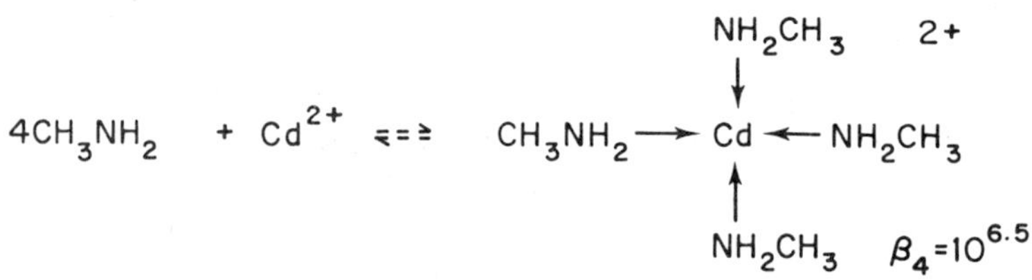

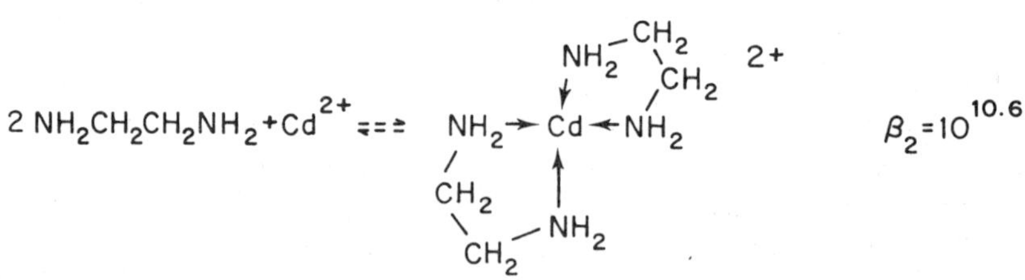

You will see that the β terms are vastly different. Our knowledge of equilibria clarifies the meaning of these values, the chelated complex is vastly more stable than the complex formed by using unidentate ligands.

As the binding sites are chemically rather similar it is worth asking why this should be so. Measurement of the enthalpy part for both of these reactions gives values in the range -55 to -58 kJ mol^{-1}, ie more or less the same, in line with the similarity in the nature of the binding sites. Consequently we might predict that the difference is related to the entropy part of the free-energy expression. We can pursue the explanation qualitatively by examining the reaction below.

Π Refer to the values for β_4 and β_2 given above and predict the general nature of the following equilibrium.

$$Cd(NH_2Me)_4^{2+} + 2en \rightleftharpoons Cden_2^{2+}\ 4\,MeNH_2$$

Our knowledge of β_4 and β_2 for the separate reactions allows us to predict that this reaction will lie strongly to the right-hand side of the equation, ie the chelating ligand has displaced the unidentate ligand.

If we look at both sides of this displacement equilibrium we can see a dipositive complex ion on each side but on the left hand side we have two molecules of ligand whereas on the right we have four. Thus the products have a greater amount of translational randomness and consequently a higher entropy. Fed back into the equation this makes for a higher negative value for ΔG° and consequently a higher value for β_n.

This chelation effect is the reason for the appearance of such a large number of polydentate ligands in coordination chemistry and in analytical chemistry. It is usually observed that chelated complexes formed by ligands which give rise to 5-membered rings are more stable than those formed by other ligands with similar binding sites. Equilibrium studies also show that the chelate effect has dropped into insignificance for ring sizes of 8 and above.

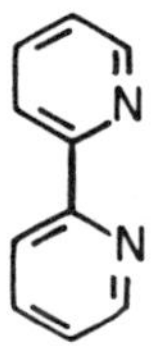

2,2'-Bipyridyl
(bipy)

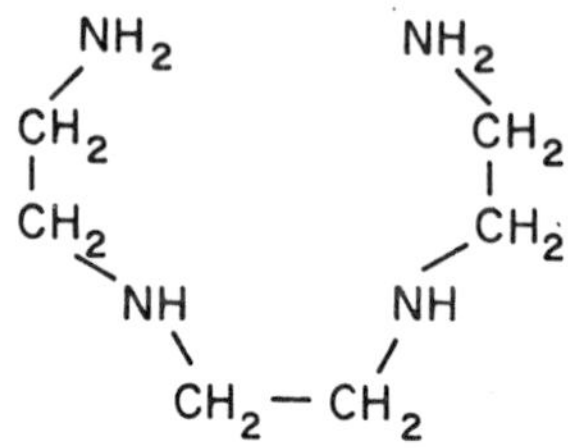

N^1,N^2-Di-(2-aminoethyl)-1,2-diaminoethane
(Triethylenetetramine or trien)

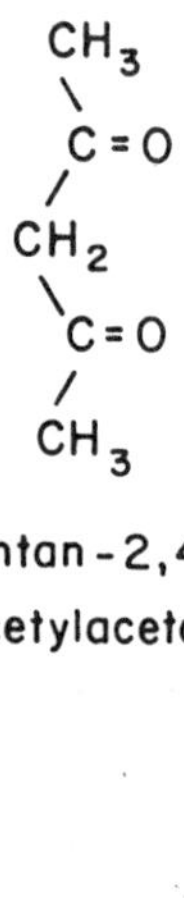

Pentan-2,4-dione
(Acetylacetone, acac)

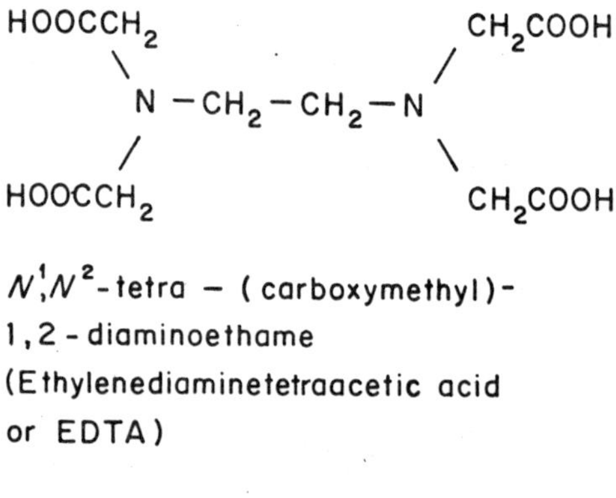

N^1,N^2-tetra – (carboxymethyl)-
1,2-diaminoethame
(Ethylenediaminetetraacetic acid
or EDTA)

Fig. 4.3b. *Some chelating ligands encountered in analytical chemistry*

The abbreviations are commonly used in texts without constant reference back to their origin so it is worth committing these to memory.

SAQ 4.3a

Which of the following ligands would you expect to form chelates?

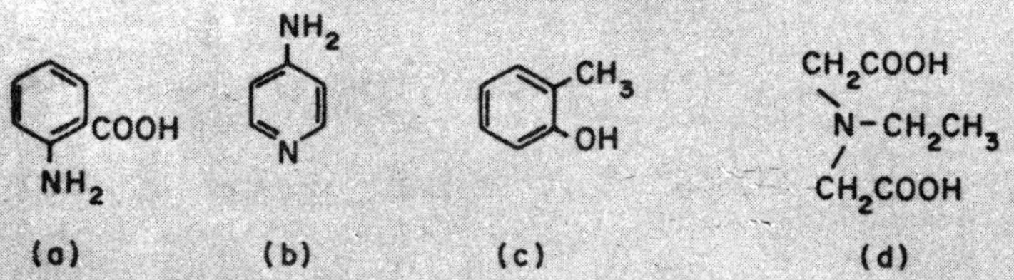

SAQ 4.3b

> **Which of the following is most correct?**
>
> **The formation constants of chelates are ...**
>
> **(*i*) ... usually greater ...**
>
> **(*ii*) ... marginally greater ...**
>
> **(*iii*) ... always much greater ...**
>
> **... than the formation constants for closely related non-chelated systems.**

4.3.1. Two Views of Stability

The equilibrium model deals with the *thermodynamic* stability of complexes. Thus the equilibrium approach to the study of coordination and its application in analytical chemistry tells us nothing about the *rates* with which equilibria are established. It is however important to be aware that there is a striking range of *kinetic* stabilities observed for complexes which are of apparently similar constitution. We use the term *kinetically stable* for reactions which are so slow that, after mixing the reactants, it may take several hours or even days for equilibrium to be established. Another term for this type of system is *inert*. In quite distinct cases reactions may be very rapid and take place in the time scale of the actual mixing of the reactants. These systems are said to be *labile*.

The empirical generalisations about this are called Taube's rules and the theoretical foundations of this behaviour are far from simple and need not concern us here. Interested readers may consult D Nicholls, *Complexes and First Row Transition Elements*, Chapter 8, Macmillan, London, 1974. The important point for us to note in the context of analytical applications is that there is a wide range of kinetic behaviour which is superimposed on our equilibrium model.

4.4. AMINOCARBOXYLIC ACIDS

Of all the various types of chelating ligand used in analytical chemistry, there is no doubt that aminopolycarboxylic acids are the most important group. The most important individual reagent in this group is ethylenediaminetetra-acetic acid, universally known as EDTA.

We shall examine some equilibrium aspects of its behaviour briefly below as an examplar for aminocarboxylic acids in general. The ligand itself has six potentially good coordinating sites, two nitrogen atoms and four carboxy groups.

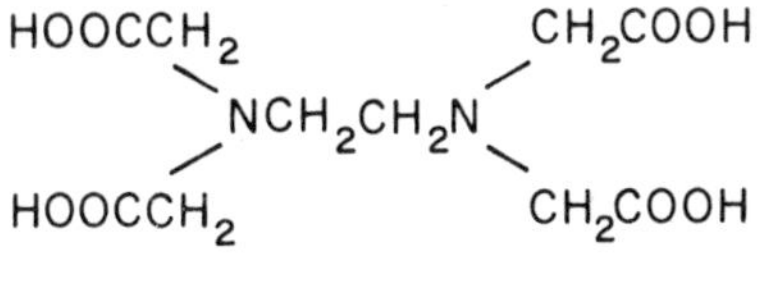

EDTA

In fact one can measure the pK values for the loss of the acidic protons and we find that the last two are much more difficult to remove than the first two;

$$pK_1 = 2.0;\ pK_2 = 2.67;\quad pK_3 = 6.2;\ pK_4 = 10.3.$$

Thus we might expect predominant behaviour as a quadridentate ligand with the potential for acting as a sexadentate ligand under favourable circumstances.

We now illustrate the tremendous importance of being able to deal with *one equilibrium* involving a quadridentate ligand *rather than four separate equilibria* when using unidentate ligands. We shall plot pM (ie – log [M]) against the addition of the ligand in a manner analogous to that for plotting pH against added acid in a neutralisation. What we observe are curves very much of the same form as the acid–base titration curves.

First we consider the stepwise formation of a hypothetical complex, ML_4, *via* the intermediaries ML, ML_2 and ML_3, with respective formation constants 10^8, 10^6, 10^4, and 10^2 for ML_4.

$$M + 4L \rightleftharpoons ML_4$$

Our equilibrium model allows us to predict that there will be no region of rapid change in pM. Curve A represents this. This is observed because there is no great difference between successive K values (Fig. 4.4a). If we compare this with a system in which the K values are segregated into two different groups with a large difference between them, we can predict that there will be two regions of fairly rapid change in pM. This is what would happen if we considered the reaction as being that of a metal-ion with two molecules of a bidentate ligand with approximate successive formation constants $K_1 = 10^{13}$, $K_2 = 10^7$.

$$M + (L\text{–}L) \rightleftharpoons M(L\text{–}L)_2, \quad [(L\text{–}L) = \text{bidentate ligand}]$$

Curve B (Fig. 4.4a) represents this case. Notice that in both the examples above $\beta = 10^{20}$.

It is of course clear that we can carry out the reaction in one step with a quadridentate ligand. In order for this to be comparable we should keep $\beta = 10^{20}$ ie as this is a one step process, $K_1 = 10^{20}$. Our equilibrium model allows us to predict a curve with a region of very rapid change in pM analogous to a strong acid-strong base type of titration curve. Curve C (Fig. 4.4a) represents this.

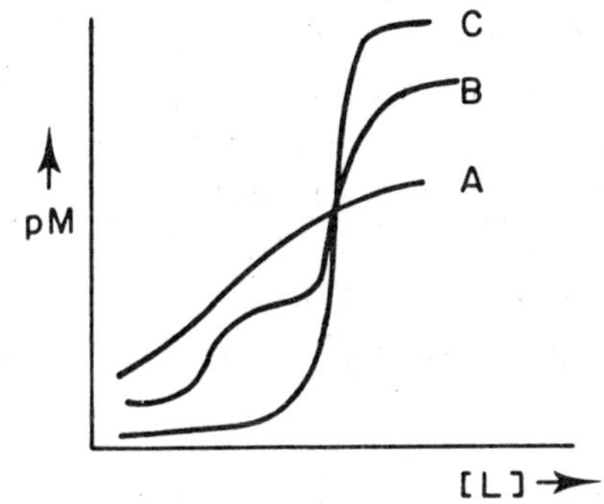

Fig. 4.4a. *Titration curves for unidentate (A), bidentate (B) and quadridentate (C) ligands added to M, see text for K values*

If you examine the structure of EDTA you will notice that it is able to form 1 : 1 metal–EDTA complexes regardless of the charge on the cation. You will also note that the coordination process is expected to lead to the formation of hydrogen ions. For example below pH 6, the predominant species in EDTA solution is a di-negative species arising from the loss of two protons. This may react as follows.

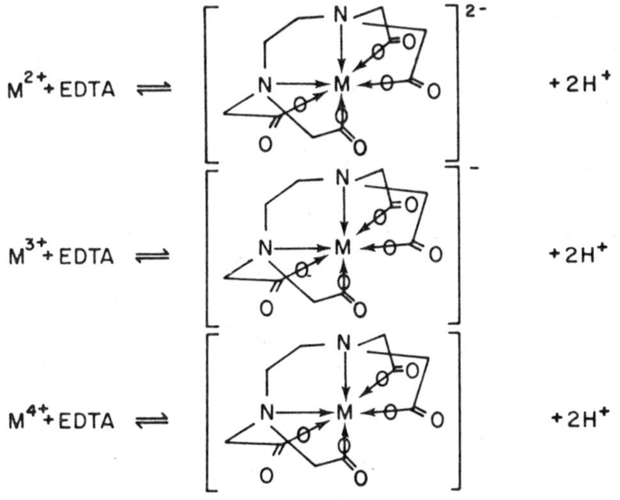

Thus we can again use our equilibrium model to predict that alkaline conditions will favour the formation of EDTA complexes. This is true, but in practice some EDTA complexes have sufficiently high formation constants to be successfully formed in acidic media. Examples are complexes with metals such as nickel ($K = 4.2 \times 10^{18}$), copper ($K = 6.3 \times 10^{18}$) and zinc ($K = 3.2 \times 10^{16}$). Metals with lower formation constants will require strongly alkaline conditions to permit titration, eg calcium ($K = 5 \times 10^{10}$) and magnesium ($K = 4.9 \times 10^{10}$).

You may have recognised that this type of behaviour offers scope for the separate titration of metals in mixtures by exploitation of adjustments in the pH. Further details of this type of application are found in Part 13.

SAQ 4.4a

Which of the curves in the figure would you expect to correspond in general form to that for the addition of the bidentate ligand 1,10-phenanthroline to a solution of a metal ion with a coordination maximum of four?

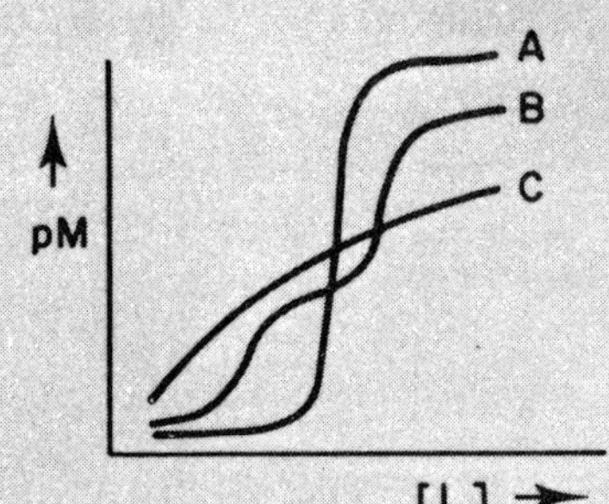

SAQ 4.4a

4.5. END-POINT DETECTION

The detection of end-points in complexometric titrations is another area in which our equilibrium model provides insight.

In titrations involving complex formation it is very common to use indicators which are essentially acid–base systems in which the anion can coordinate with the metal atom. An obvious requirement is intense colour in the species involved. Usually we have the displacement of the indicator species, In, from a *weakly* bound chelate MIn to form the more strongly bound M(EDTA) complex.

$$\text{MIn} + \text{EDTA} \rightleftharpoons \text{M(EDTA)} + \text{In}$$

Typical indicators are Eriochrome Black T, and Pyrocatechol Violet; both are shown below with the predominant colour in different pH ranges corresponding to successive ionisations.

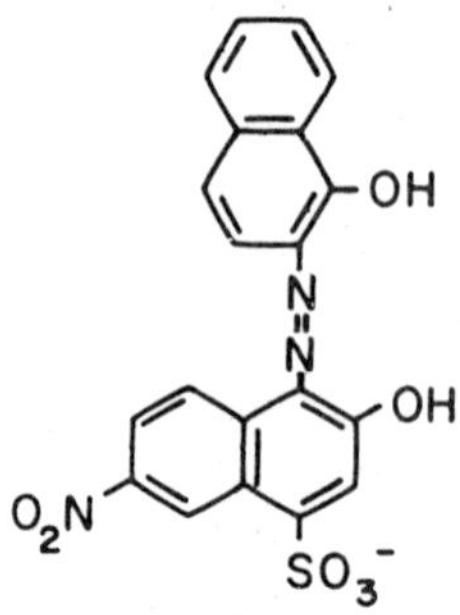

pH<5.5
Red

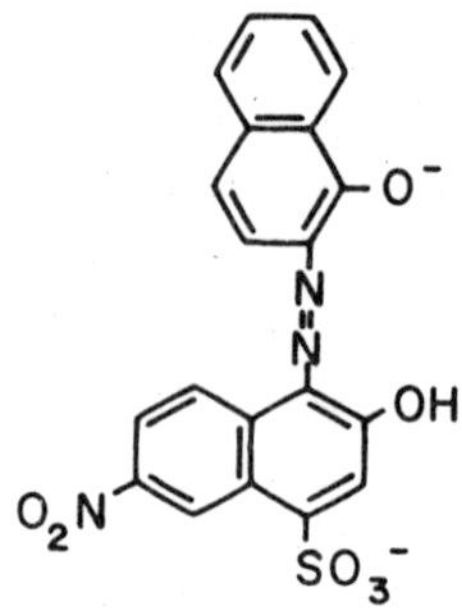

pH7 - 11
Blue

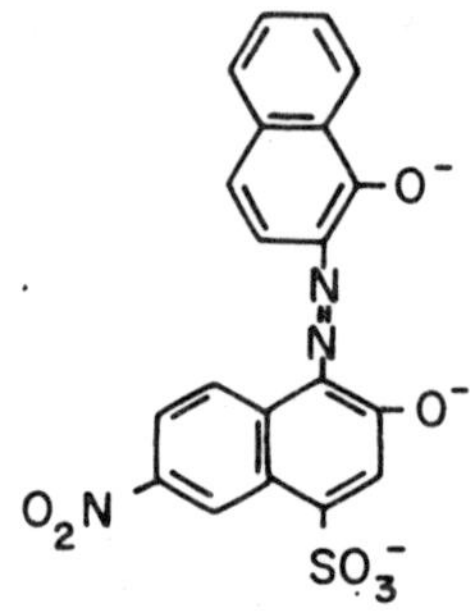

pH > 11.5
Orange

Eriochrome Black T

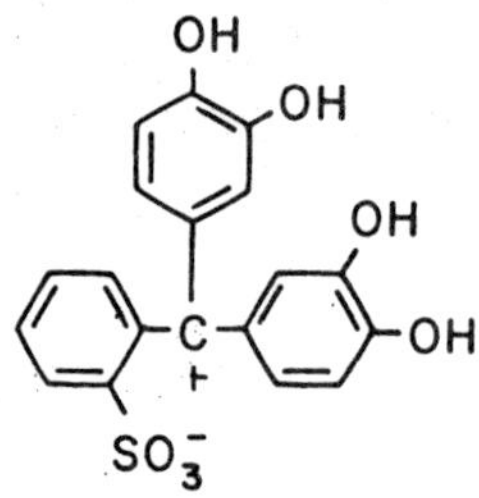

pH<1.5
Red

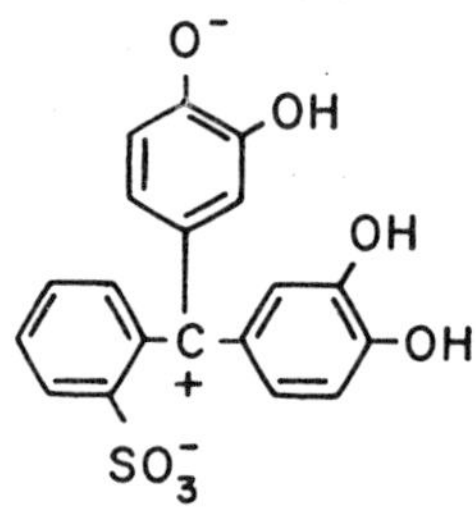

pH 2-6
Yellow

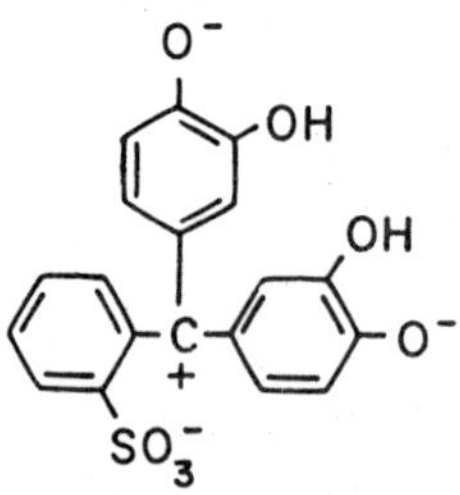

pH 7
Violet

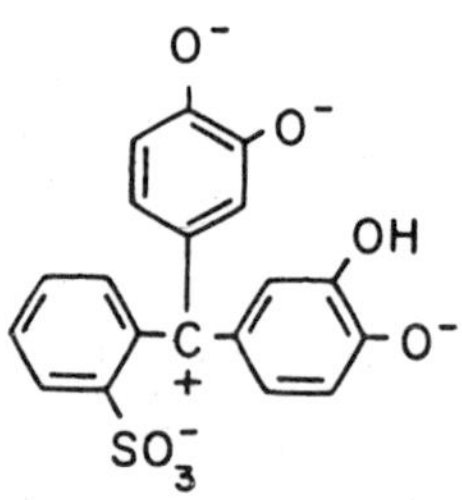

pH>10
Blue

Pyrocatechol Violet

Fig. 4.5a. *Indicators for complexometric titrations*

The metal-ion complexes with Eriochrome Black T and with Pyrocatechol Violet are coloured red and blue–green respectively. Thus we can see on colour grounds alone that alkaline conditions might suit Eriochrome Black T (red to blue) whereas acidic conditions suit Pyrocatechol Violet (blue–green to yellow).

Our equilibrium model however should advise a little caution before proceeding. We have seen that the indicator action is in fact a ligand displacement reaction. Clearly for this to be satisfactory in the laboratory the formation constant for the EDTA complex must be significantly greater than that for the indicator complex otherwise premature end-points would be observed. However, there is also the operational requirement that the formation constant for the indicator complex be large enough for it to be formed in sufficient concentration to be detectable. We usually look for a minimum difference of 10^2 in the formation constants of the indicator complex and the EDTA complex; in practice this is not difficult to achieve. The equilibrium nature of these reactions also tells us that it is desirable to use indicators for which the complex MIn is much more intensely coloured than the free indicator, and that it is desirable to work with the minimum amount of indicator commensurate with end-point observation.

Objectives

Now that you have completed this Part you will be able to:

- identify the process of complex formation and the function of ligands,

- recognise ligands capable of chelate formation,

- identify common analytical ligands,

- describe the stepwise formation of complexes and the associated formation constants,

- interpret the significance of the relative magnitudes of formation constants,

- draw typical titration curves for complex-formation reactions,
- relate the behaviour of complexometric indicators to the equilibrium model.

5. Redox Systems

This Part deals with the application of the equilibrium model to redox systems with special reference to redox titrations. Some basic electrochemical aspects such as electrochemical cells, the concept of the half-reaction, and oxidation numbers are described along with the sign convention for electrode potentials. The relationship between system potential and the relevant concentrations (the Nernst equation) is developed and applied to the calculation of theoretical titration curves and equivalence-points. The behaviour of mixed redox titrations is described and this is extended to include the use of redox indicators.

5.1. ELECTRODE POTENTIALS

5.1.1. Basic Electrochemical Aspects

The term *redox* is used to describe a system in which one component is oxidised and at the same time another component is reduced. If we recall the meaning of oxidation and reduction in terms of electrons the redox process becomes clearer. Oxidation of a substance, which may be an element or a compound, involves the loss of electrons by it. Conversely reduction of a substance involves the gain of electrons by it. Thus in a redox system we are essentially concerned with *electron transfer*. Note that the electron transfer is from the substance which is oxidised to the substance which is reduced.

You will find that chemists frequently talk about *oxidising agents* (*or oxidants*) and *reducing agents* (*or reductants*). You should be careful to relate these terms accurately to the oxidation and reduction of substances. It is the oxidant which is reduced, thus we can accurately state that the substance undergoing oxidation is in fact the reductant. This apparent complication springs from the early view that one reagent was in some sense the active constituent and the other reagent was simply a passive partner in the reaction. We should note, however, that this is not so. The so-called oxidising agent simply starts out in a highly oxidised form and finishes in a less oxidised form, ie it undergoes reduction during the reaction.

Another minor problem springing from general usage by chemists is the apparent view that the terms *oxidant* and *reductant* are absolute. This can lead to the belief that a particular compound is always an oxidant (or reductant as the case may be). This is not true, some materials are oxidants in the presence of certain reagents and reductants in the presence of others. For example, in the reaction of iodine with iodate (sometimes used as part of a sequence in the volumetric determination of iron) we see the oxidation of iodine, ie iodine is the reductant.

$$IO_3^- + 2I_2 + 6H^+ \rightleftharpoons 5I^+ + 3H_2O$$

On the other hand, iodine may undergo reaction with arsenic(III) oxide (arsenious oxide), eg in the standardisation of solutions, in which we can see that the iodine is the oxidant and is itself reduced to iodide.

$$As_2O_3 + 2I_2 + 2H_2O \rightleftharpoons As_2O_5 + 4H^+ + 4I^-$$

Thus the terms oxidant and reductant are to be used with caution, recognising that they refer to some specific process and serve *as a guide only* for general observations of behaviour.

If we take the reaction of iron(II) ions and permanganate ions as a typical oxidation-reduction equilibrium applied in analysis (very strongly to the right-hand side), we can separate the part undergoing oxidation from that undergoing reduction.

$$MnO_4^- + 5Fe^{2+} + 8H^+ \rightleftharpoons Mn^{2+} + 5Fe^{3+} + 4H_2O$$

Clearly you will see that the part undergoing oxidation involves the iron; Fe^{2+} is oxidised to Fe^{3+}. We could equally well say that in this system Fe^{2+} is reducing permanganate ion. The component undergoing reduction, ie doing the oxidation is the permanganate ion, MnO_4^-; this is reduced to the manganese(II) ion (manganous ion), Mn^{2+}. Note that here we have Mn(+7), ie highly oxidised manganese, becoming Mn(+2), ie less oxidised manganese. This does not of course exclude the possibility of further reduction, eg by electrolysis, to give Mn(0). Although this division of the reaction equation into two parts seems artificial, it allows us to develop a very useful model for dealing with redox equilibria.

5.1.2. Oxidation Numbers

Although in principle oxidation numbers can be determined in a species by assigning bonding electrons in the Lewis structures, in practice it is simpler to apply the following four rather arbitrary rules.

(*a*) The oxidation number of an element in the elemental state is 0, eg Br_2, H_2, P_4; oxidation numbers all 0.

(*b*) The oxidation number of an element in a monatomic ion is equal to the charge on the ion, eg Ca^{2+} is Ca(+2), Na^+ is Na(+1), I^- is I(−1).

(*c*) Certain general propositions apply, *viz.* group I metals are always +1 in their compounds, group II metals are always +2 in their compounds, oxygen is always −2, fluorine is always −1, and hydrogen is always +1 except in metal hydrides such as NaH and $LiAlH_4$.

(*d*) The sum of the oxidation numbers for the elements which constitute a species is equal to the charge on the species (zero for a neutral species).

You will no doubt agree that a more quantitative way of referring to redox processes is desirable. However before proceeding, try the following SAQ's to clarify your grasp of the terminology.

SAQ 5.1a

What is the oxidation state of the following?

(*i*) Mn in MnO_4^-,

(*ii*) Cu in $[Cu(NH_3)_4]^{2+}$,

(*iii*) Se in SeO_3^{2-},

(*iv*) Cr in $Cr_2O_7^{2-}$,

(*v*) I in I_2,

(*vi*) I in HI,

(*vii*) N in N_2H_4.

SAQ 5.1b

Which is the reactant undergoing *oxidation* in the following equilibria?

(*i*) $SO_3^{2-} + I_2 + H_2O \rightleftharpoons SO_4^{2-} + 2H^+ + 2I^-$

(*ii*) $12H^+ + 4MnO_4^- + 5Sb_2O_3 \rightleftharpoons 4Mn^{2+} + 6H_2O + 5Sb_2O_5$

(*iii*) $2Cu^{2+} + 4I^- \rightleftharpoons 2CuI + I_2$

(*iv*) $2MnO_4^- + 10I^- + 16H^+ \rightleftharpoons 2Mn^{2+} + 5I_2 + 8H_2O$

SAQ 5.1c

For the following equilibria which reactant is the oxidant?

(*i*) $10\,Cl^- + 2\,BrO_3^- + 12\,H^+ \rightleftharpoons 5\,Cl_2 + Br_2 + 6\,H_2O$

(*ii*) $2\,S_2O_3^{2-} + I_2 \rightleftharpoons S_4O_6^{2-} + 2\,I^-$

(*iii*) $5\,V^{2+} + 3\,MnO_4^- + 24\,H^+ \rightleftharpoons 5\,V^{5+} + 3\,Mn^{2+} + 12\,H_2O$

(*iv*) $Fe^{3+} + Ti^{3+} \rightleftharpoons Fe^{2+} + Ti^{4+}$

If we refer to the equation for the reaction of iron(+2) and permanganate ion, we can see that it is possible to write separate equations which deal with only half of the original reaction, ie we can write out the oxidising part separately from the reducing part. In order to balance these equations for the half-reactions we may need to add water (to balance the oxygen atoms), and H^+ (to balance the hydrogen atoms). This is then followed by adding electrons to balance the charges on either side.

$$Fe^{2+} = Fe^{3+} + e; \qquad \text{an oxidation}$$

$$MnO_4^- + 8H^+ + 5e = Mn^{2+} + 4H_2O; \quad \text{a reduction}$$

If you examine these half-reactions you will notice that the number of electrons appearing in the equation corresponds to the change in oxidation number of the 'central' element. You will also note that as the electrons appear on different sides of the equations, they will cancel if the equations are multiplied by an appropriate number then added. In the example above you will see that we need five times the oxidation equation to balance one of the reduction equation. The combination gives, of course, our original equation but you will now appreciate that what has in fact happened is that electrons have been transferred *from* the iron *to* the manganese centre.

SAQ 5.1d

Write balanced equations for the half-reactions which represent the changes below and use them to produce balanced equations for the reactions which follow (*i*) to (*v*).

$ClO_3^- \rightarrow Cl^-$

$RNO_2 \rightarrow RNH_2$

$S^{2-} \rightarrow SO_4^-$

$I^- \rightarrow I_2$

$Ti^{3+} \rightarrow Ti^{4+}$

⟶

SAQ 5.1d (cont.)

$OBr^- \rightarrow Br^-$

$H_2O_2 \rightarrow H_2O$

$IO_3^- \rightarrow I_2$

$As_2O_3 \rightarrow As_2O_5$

(*i*) The reaction of chlorate ions and iodide ions in acidic media.

(*ii*) Reduction of nitro compounds by using titanium(+3).

(*iii*) The reaction of sulphide ions with bromate(I) ions (hypobromite).

(*iv*) The oxidation of arsenic(III) oxide by hydrogen peroxide.

(*v*) The reaction of iodate ions with iodide ions.

Try the next SAQ to reinforce some of these ideas.

SAQ 5.1e

Mark the following statements about redox reactions as 'true' (T) or 'false' (F).

(*i*) Redox reactions are either oxidising or reducing reactions depending on the prevailing conditions.

(T) (F)

(*ii*) If the oxidant is added to the other reagent the process is always described as an oxidation and reduction does not take place.

(T) (F)

(*iii*) Redox reactions involve both a reduction and an oxidation process occurring simultaneously.

(T) (F)

(*iv*) In redox processes the oxidation number of the oxidant increases.

(T) (F)

(*v*) In redox processes the electrons are transferred to the reductant from the oxidant.

(T) (F)

(*vi*) The change in oxidation numbers of oxidant and reductant arises from electron transfer.

(T) (F)

(*vii*) The oxidising agent is the material which gains electrons during a redox process.

(T) (F)

⟶

SAQ 5.1e (cont.)

(*viii*) The change in the oxidation number of the oxidant must be equal to that of the reductant.

(T) (F)

(*ix*) Strong oxidising agents generally contain central elements with high positive oxidation numbers.

(T) (F)

An interesting feature of this seemingly artificial separation of the halves of the reaction is that they correspond to the anodic and cathodic parts of a galvanic cell.

Although our principal interest is in equilibrium aspects of redox processes, it will be useful here to reiterate features of galvanic cells which you may have met before. Many of us are familiar with the fact that dipping ferrous metals or zinc-based alloys into copper(II) sulphate solution leads to 'plating-out' of copper onto the metal surface. This is clearly a reduction of copper(+2) ions to copper(0) metal.

The half reaction is:

$$Cu^{2+} + 2e = Cu^0$$

We also know that something must be undergoing oxidation to produce the required electrons. Zinc metal is the source of these electrons, if we are examining the example of a zinc alloy in copper sulphate solution. You will find that zinc is the most common example used in introductory texts dealing with this subject. The appropriate half-reaction is:

$$Zn^0 = Zn^{2+} + 2e$$

We can visualise this reaction as the transfer of electrons from the zinc to the copper. We all know, of course that the movement of electrons in wires is the same thing as the passage of an electric current. The question we wish to ask is, can this electron transfer between the zinc and the copper take place *via* the intermediacy of an electrical connection (eg by a wire)?

It turns out that the answer is yes.

The reaction of the copper ions and zinc metal does not require the the direct contact of these species, but it does require a means by which electrons can be exchanged and also a means by which ions can flow. The diagram of a simple galvanic cell illustrates this (Fig. 5.1a). It is the potential difference between the electrodes which represents the driving force of the chemical reaction, ie the redox process below.

$$Cu^{2+} + Zn^0 \rightleftharpoons Cu^0 + Zn^{2+}$$

Our interest is to attempt to quantify this driving force and relate it to the equilibrium model which we have already developed. We can see that the reaction is made up of two parts (the oxidation part and the reduction part) consequently it should come as no surprise that we can approach the idea of the driving force for the whole process by examining that for each half. That is, we wish to examine the potential associated with each half-reaction. As for any measurement, we need some standard against which we can make comparisons. The equation for the reaction reveals an immediate problem, part of the process is a reduction and part is an oxidation. Further we have not established any standard against which we can measure the differences in potential.

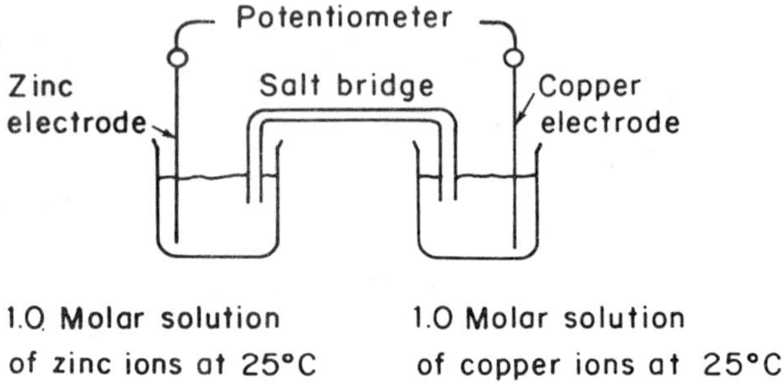

Fig. 5.1a. *The galvanic cell represented by* $Zn/Zn^{2+}_{(aq)}/Cu^{2+}_{(aq)}/Cu$; *this is also known as the Daniell cell*

5.1.3. Sign Conventions for Potentials

You will have noticed that the scope for confusion in the absence of absolutes and standards is enormous. It is necessary that we all agree on the conventions and signs used for the writing of half-reactions, for the signs of the associated potentials, and for the standard against which all our measurements are to be made.

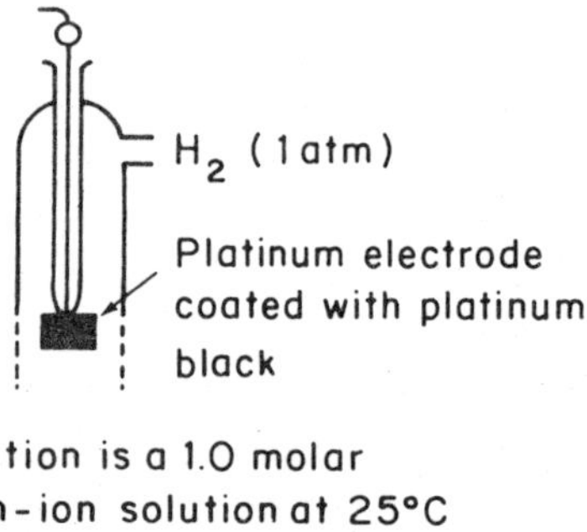

Fig 5.1b. *The standard hydrogen electrode,* $Pt_{(s)}H_{2(g)}/H^{+}_{(aq)}$

The scientific world has selected the *standard hydrogen electrode* as the electrode to which all other electrode potentials should be referred. This is an electrode for which the hydrogen-ion activity is unity and in which there is a partial pressure of one atmosphere of hydrogen gas (Fig. 5.1b). Thus if we write the half-reaction as a *reduction* we have

$$H^+ + e \rightleftharpoons \tfrac{1}{2}H_2 \; (a_{H^+} = 1, p_{H_2} = 1 \text{ atmosphere}) \; E^o = 0$$

Note that as this is the standard against which all other potentials will be measured: it is arbitrarily given the value of zero.

Another important and essentially arbitrary convention (agreed at the IUPAC Stockholm meeting in 1953) is that the term *electrode potential* should be used *exclusively* for half-reactions written as *reduction equations*, (ie electrons on the left-hand side). The sign of the potential for a half reaction is the actual sign the electrode in question would have if coupled to the standard hydrogen electrode to form a galvanic cell. You are particularly warned to be wary in the use of data from older books dealing with electrode potentials, especially as one of the standard works for many years expresses half-reactions as oxidations. (W.M.Latimer, *The Oxidation States of the Elements and their Potentials in Aqueous Solutions*, Prentice-Hall, New York, 1952).

Thus the IUPAC approach to the example of the half-reaction involving zinc is to write it as follows.

$$Zn^{2+} + 2e \rightleftharpoons Zn_{(s)} \qquad E^o = -0.763 \text{ V}$$

That is, in a galvanic cell consisting of a hydrogen electrode and a zinc electrode, the zinc will dissolve and become zinc(+2) ions and hydrogen ions will be discharged as hydrogen gas. The overall driving force as measured by the cell potential will be given by subtracting the *lower* half-reaction potential (in this case −0.763 V) from the *higher* half-reaction potential (in this case 0.0 V). Thus the cell potential is 0.0 − (−0.763) = +0.763 V. Remember that if you get negative numbers after this type of manipulation it indicates that you have got the half-reactions the wrong way round. Potential differences are in fact differences and must always be positive.

Latimer would have written this as:

$$Zn_{(s)} \rightleftharpoons Zn^{2+} + 2e \qquad E^{o} = +0.763 \text{ V}$$

You will see that it is really the same data, the equation is reversed and so is the sign of E^{o}.

It is a good idea to commit to memory the fact that in the IUPAC convention the half-reactions for the strongest oxidants have positive values of E^{o}.

5.1.4. Concentration and Redox Processes

Our knowledge of acid-base equilibrium systems leads us to expect the *concentration or activity of chemical species* to play some part in determining the overall behaviour of the system.You will recall that the same was true for the behaviour of coordination compounds. Indeed it comes as no surprise to find the concentrations of some of the species involved playing an important part in the behaviour of redox systems.

This is expressed through an effect on the driving force, ie an influence on the potential for the half-reaction expressed in the equation below.

$$E = E^{o} - \frac{RT}{nF} \ln \frac{[\text{reduced form}]}{[\text{oxidised form}]}$$

E = potential for the half-reaction, V,

E^{o} = a standard electrode potential (characteristic of the half-reaction), V,

R = the molar gas constant, 8.314 volt-coulombs K^{-1} mol^{-1},

T = absolute temperature, K,

n = number of electrons transferred,

F = the Faraday, 96500 coulombs,

ln = natural logarithm, ie 2.303 $\log_{10}$.

Expressed using logs to base 10, this becomes for 298 K;

$$E = E^{\circ} - \frac{0.0591}{n} \log \frac{[\text{reduced form}]}{[\text{oxidised form}]}$$

These are both forms of an expression known as the Nernst equation.

Strictly the equation should contain activities rather than concentrations. However, as we saw in earlier discussion of equilibria it is quite common for many applications to use concentrations rather than activities and accept the small errors thereby introduced. You will have noticed that yet again we have had to introduce a 'standard', in this case the *standard electrode potential*.

It is easy to see where this comes from if we consider the Nernst equation for half-reactions in which the activity of the oxidised form and that of reduced form are both unity. When these terms are equal (here equal to 1) then the whole of the activity quotient becomes 1, hence the logarithmic part becomes zero.

Thus E becomes equal to the half-cell potential when all reactants and products are at unit activity, or when the activity ratio is unity. As the electrode potentials are temperature-dependent the additional restriction of a constant temperature of 25 °C, ie 298 K, is built into the definition of E°. A word of caution here, ... although the activity quotient looks like an equilibrium constant, it is not one; the systems are not in a state of equilibrium.

A few standard electrode potentials are given in the table below (Fig. 5.1c) along with the equations for the relevant half-reactions.

You can gain a qualitative view of the importance of the Nernst equation by looking at some examples from the table, eg Eq. (2).

Note that the activity quotient will have a term in $[H^+]$.

$$E = E^{\circ} - \frac{0.0591}{5} \log \frac{[Mn^{2+}]}{[MnO_4^-][H^+]^8} \quad \text{at 298 K}$$

You can see that the potential of the system varies with the pH.

		E°, V (25 °C, unit activity)
(1)	$MnO_4^- + 4H^+ + 3e = MnO_2 + 2H_2O$	+1.69
(2)	$MnO_4^- + 8H^+ + 5e = Mn^{2+} + 4H_2O$	+1.51
(3)	$Ce^{4+} + e = Ce^{3+}$	+1.45 (M H_2SO_4)
(4)	$2BrO_3^- + 12H^+ + 10e = Br_2 + 6H_2O$	+1.52
(5)	$Cr_2O_7^{2-} + 14H^+ + 6e = 2Cr^{3+} + 7H_2O$	+1.33
(6)	$VO_4^{2-} + 6H^+ + 2e = VO^{2+} + 3H_2O$	+1.2
(7)	$2IO_3^- + 12H^+ + 10e = I_2 + 6H_2O$	+1.19
(8)	$Sn^{4+} + 2e = Sn^{2+}$	+0.14
(9)	$Br_2 + 2e = 2Br^-$	+1.065
(10)	$Fe^{3+} + e = Fe^{2+}$	+0.771
(11)	$2H^+ + 2e = H_{2(g)}$	0.0
(12)	$Cd^{2+} + 2e = Cd_{(s)}$	−0.403
(13)	$Cr^{3+} + e = Cr^{2+}$	−0.38
(14)	$S + 2e = S^{2-}$	−0.7
(15)	$Zn^{2+} + 2e = Zn_{(s)}$	−0.763

Fig. 5.1c. *Selected values of standard electrode potentials*

Similar reasoning will apply to Eqs. (1), (4), (5), (6), and (7).

For Eq. (12) the activity quotient reduces to the reciprocal of the cadmium-ion concentration.

$$E = E^{\circ} - \frac{0.0591}{2} \log \frac{[Cd]_{(s)}}{[Cd^{2+}]}$$

Here we note that the activity of a solid is unity by definition.

Hence the potential varies linearly with the log of the reciprocal of the cadmium-ion concentration. Similar reasoning will apply to Eqs. (14) and (15). For Eq. (14) it is the sulphur which has unit activity hence E will vary linearly with the log of the sulphide-ion concentration, not the reciprocal.

For Eq. (10) the activity quotient reduces to a *ratio* of ionic concentrations.

$$E = E^{\circ} - \frac{0.0591}{2} \log \frac{[Fe^{2+}]}{[Fe^{3+}]}$$

It is also useful to explore the meaning of these potentials in terms of the actual reactions rather than our half-reactions.

Let us take the possible reaction of cerium(+4) with iron(+2). Reference to the table shows that for the cerium part, ie Eq. (3), E° is +1.45 V. For the iron part, Eq. (9), E° is only 0.78 V. Thus cerium(+4) will oxidise iron(+2), and the driving force for the reaction is measured as the difference between the two E° values, ie $1.45 - 0.78 = 0.67$ V (*at unit activity quotient*).

Note that what we are saying is that for the pair of half-reactions being compared, the half-reaction with the *more positive* value of E° is written in the direction specified in the form used to give standard electrode potentials. The chemical reaction for the other half-reaction then, has the reverse of the normal direction for half-reactions. That is, the half-reaction with the more positive value of E° proceeds as a reduction, and the other half-reaction proceeds as an oxidation.

SAQ 5.1f

By using the data in Fig. 5.1c (E° values), state whether the following statements are true (T) or false (F), when the constituents are at unit activity.

(*i*) Fe^{3+} is expected to oxidise Br^-.

(T) (F)

(*ii*) MnO_4^- is a more powerful oxidant than $Cr_2O_7^{2-}$.

(T) (F)

(*iii*) IO_3^- is capable of oxidising chromium(+2) to chromium(+3) but not to chromium(+6).

(T) (F)

(*iv*) Metallic zinc should be insoluble in solutions containing ferric ions.

(T) (F)

(*v*) Potassium bromate will not be oxidised by potassium dichromate.

(T) (F)

(*vi*) Cadmium metal placed in a solution of zinc ions should lead to the plating-out of zinc metal.

(T) (F)

(*vii*) Although tin(+2) is properly described as usually having reducing properties it will not reduce bromine to bromide.

(T) (F)

SAQ 5.1f

Although the standard electrode potentials from the table may be used for a quick qualitative assessment of the likelihood of a reaction proceeding, *do not forget* that the actual values of E depend on concentrations, *and* that the concentrations will normally be changing.

Thus we can picture the reaction as proceeding towards an equilibrium at which E(oxidation reaction) = E(reduction reaction).

We can use the Nernst equation to calculate values for E in a quite straightforward manner.

Π What is the potential at 25 °C for a half-cell made up of a cadmium electrode in contact with 0.0200M cadmium ion solution?

$$Cd^{2+} + 2e \rightleftharpoons Cd_{(s)};\ E^{o} = -0.403\ V.$$

$$E = E^{o} - \frac{0.0591}{n} \log \frac{[\text{reduced form}]}{[\text{oxidised form}]}$$

As $[Cd_{(s)}] = 1$

$$E = E^{o} - \frac{0.0591}{2} \log \frac{[1]}{[Cd^{2+}]}$$

$$= -0.403 - 0.0296\,[\log(1/0.02)]$$

$$= -0.453 \text{ V.}$$

Now try the following SAQs.

SAQ 5.1g

For an electrochemical cell in which the cell reaction is;

$$Zn^{o}_{(s)} + Cu^{2+} \rightleftharpoons Cu^{o}_{(s)} + Zn^{2+}$$

predict:

(*i*) the effect of increasing the concentration of Cu^{2+} ions on the cell voltage,

(*ii*) the effect of increasing the concentration of Zn^{2+} ions on the cell voltage,

(*iii*) the effect of increasing the size of the zinc electrode on the cell voltage.

SAQ 5.1h

For an electrochemical cell in which the cell reaction is:

$$H_{2(g)} + 2\,AgCl_{(s)} \rightleftharpoons 2\,Ag_{(s)} + 2\,H^{+}_{(aq)} + 2\,Cl^{-}_{(aq)}$$

what is the effect of the following factors on the cell potential?

(*i*) Decreasing the pH,

(*ii*) Increasing the pressure of H_2,

(*iii*) Decreasing the amount of silver chloride present,

(*iv*) Increasing the molarity with respect to chloride ions.

Try making the prediction first by using the Nernst equation, then see whether Le Chatelier type reasoning leads to the same conclusion.

5.2. TITRATION CURVES FOR REDOX REACTIONS

Now that you are familiar with the general qualitative approach to these redox systems we can try some quantitative examples.

This simply involves the insertion of the relevant activities into the Nernst equation. Remember however that usually we shall use concentrations in place of activities and accept the small errors involved.

SAQ 5.2a

Calculate the potential for the half-cell consisting of 0.0100*M* KBr in the presence of liquid bromine. (In practice a platinum electrode is used for this system to avoid complications due to oxidation of other electrode materials). Note that in this example the aqueous phase is always saturated with bromine.

$$Br_{2(l)} + 2e \rightleftharpoons 2\,Br^- \qquad E^\circ = 1.065\ V$$

Our purpose in pursuing this description of redox behaviour has been to enable us to develop a model for the changes which take place during redox titrations.

Remember that in an acid–base titration we have essentially a set of different equilibrium systems (as the titration proceeds) and that the end-point is marked by a predictable region of rapid change in the pH of the solution.

In a redox titration we are looking for an end-point in a region of rapid change in the potential of the system. The way in which we detect the end-point will be described later. We can now relate the potential of a chemical system to the changes which occur during the redox titration. Let us examine the titration of 0.1*M* iron(II) sulphate with 0.1*M*-cerium(IV) sulphate. A look at the table of E° values tells us that qualitatively we expect the reaction to proceed and that the iron(+2) is oxidised to iron(+3) as cerium(+4) is reduced to cerium(+3).

However as the data in the table refer to standard conditions we need to be careful about calculating actual potentials for the concentrations in our titration.

We will consider 100 cm^3 of iron solution initially present.

(*a*) After the addition of 10.0 cm^3 of cerium(+4) solution, the ratio of the $[Fe^{2+}]$ to $[Fe^{3+}]$ is given by

$$[Fe^{2+}]/[Fe^{3+}] = (100 - 10.0)/10.0$$

$$= 9.0$$

therefore

$$E = 0.78 - 0.0591 \log 9.0$$

$$= 0.72 \text{ V}.$$

(*b*) Just short of the end-point, we have added 99.0 cm^3 of cerium(+4) solution.

The $[Fe^{2+}]/[Fe^{3+}]$ ratio is now (100 − 99)/99, ie approximately 0.01, therefore $$E = 0.78 - 0.0591 \log 0.01$$

$$= 0.9 \text{ V.}$$

(*c*) The titration is now fractionally past the end-point. There is now an excess of cerium(+4) so we will use the value of E° for Ce^{4+}/Ce^{3+} in the calculations.

After the addition of 101 cm^3 of cerium(+4) solution, the ratio

$$\frac{[Ce^{3+}](\text{formed})}{[Ce^{4+}](\text{left})} = \frac{100}{101 - 100} = 100$$

therefore $$E = 1.45 - 0.0591 \log 100$$

$$= 1.33 \text{ V.}$$

(*d*) At 10 cm^3 of cerium(+4) beyond the end-point.

$$\frac{[Ce^{3+}](\text{formed})}{[Ce^{4+}](\text{left})} = \frac{100}{110 - 100} = 10$$

therefore $$E = 1.45 - 0.0591 \log 10$$

$$= 1.39 \text{ V.}$$

You can see that there is a rapid change in the value of E as we go through the end-point, in fact the titration curve has the same general form as that of an acid–base titration.

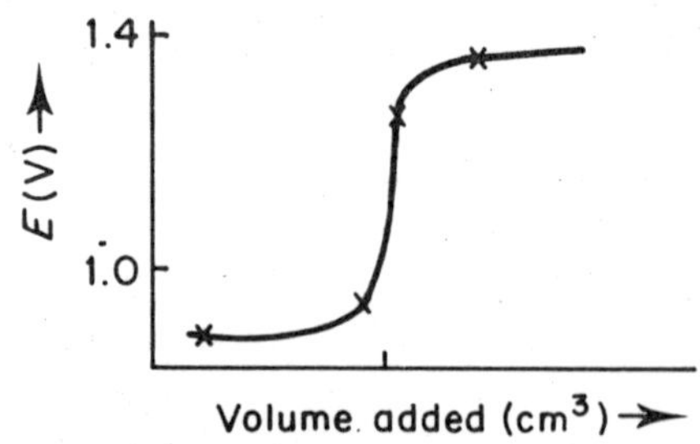

Fig. 5.2a. *Titration curve for the addition of 0.1M cerium (IV) to 0.1M iron (II)*

We can, of course, calculate the value of E exactly at the end-point. However the calculation for exact equivalence is a little more complex than the approximate method which we have just used for the approach to the end point.

For the appropriate half-reactions we have:

$$E_{\text{cerium}} = E^{\text{o}}_{\text{cerium}} - 0.0591 \log \frac{[Ce^{3+}]}{[Ce^{4+}]}$$

and

$$E_{\text{iron}} = E^{\text{o}}_{\text{iron}} - 0.0591 \log \frac{[Fe^{2+}]}{[Fe^{3+}]}$$

At the end-point the system is in equilibrium therefore:

$$E_{\text{system}} = E_{\text{cerium}} = E_{\text{iron}}$$

Thus by adding the two equations we get

$$2E_{\text{system}} = E_{\text{cerium}} + E_{\text{iron}}$$

$$= E^{\text{o}}_{\text{cerium}} + E^{\text{o}}_{\text{iron}} - 0.0591 \log \frac{[Ce^{3+}][Fe^{2+}]}{[Ce^{4+}][Fe^{3+}]}$$

But the stoichiometry of the system requires the following relationships to hold at equivalence;

$$[Ce^{3+}] = [Fe^{3+}] \quad \text{and} \quad [Ce^{4+}] = [Fe^{2+}]$$

Hence

$$\frac{[Ce^{3+}]}{[Fe^{3+}]} = 1 \quad \text{and} \quad \frac{[Fe^{2+}]}{[Ce^{4+}]} = 1$$

Hence the term within the log reduces to 1 and the log term therefore becomes zero.

Thus the potential at the equivalence point becomes

$$E_{system} = (E^o_{cerium} + E^o_{iron})/2$$

$$= (1.45 + 0.78)/2 = 1.11 \text{ V}.$$

The calculation we have just performed has a term which is reminiscent of an equilibrium constant. In fact if we write a more general redox equation we can use this approach to show how equilibrium constants may be determined from measurements of electrode potentials.

We take two species labelled (1) and (2). These have the half-reactions shown below with the coefficients (*a*) and (*b*) being required to produce the same number of electrons in each change, ie *n*.

$$a\text{OX}_{(1)} + ne \rightleftarrows a\text{RED}_{(1)}$$

$$b\text{OX}_{(2)} + ne \rightleftarrows b\text{RED}_{(2)}$$

The overall equilibrium is:

$$a\text{RED}_{(1)} + b\text{OX}_{(2)} \rightleftarrows a\text{OX}_{(1)} + b\text{RED}_{(2)}$$

so we now have an expression for K_{eq},

$$K_{eq} = \frac{[\text{OX}(1)]^a\,[\text{RED}(2)]^b}{[\text{OX}(2)]^b\,[\text{RED}(1)]^a}$$

Refer back to the worked example and convince yourself that the term within the log part is in fact K_{eq}.

The general expression is;

$$E^o_{(2)} - E^o_{(1)} = \frac{0.0591}{n} \log K_{eq}$$

or alternatively:

$$\log K_{eq} = \frac{n\{E^o_{(2)} - E^o_{(1)}\}}{0.0591}$$

Π For the iron(+2)/iron(+3)-cerium(+3)/cerium(+4) system used in the text above, calculate the equilibrium constant for the reaction.

Note that the reaction is:

$$Fe^{2+} + Ce^{4+} \rightleftharpoons Fe^{3+} + Ce^{3+}$$

and that here $n = 1$.

$$\text{Hence } \log K_{eq} = \frac{1}{0.0591}(1.45 - 0.78)$$

$$= 11.337$$

$$\text{and } K_{eq} = 2.1 \times 10^{11}$$

SAQ 5.2b

Calculate the equilibrium constant for the reaction between iron(+2) and manganate(+7) (the permanganate ion).

MnO_4^-/Mn^{2+}, $E^o = +1.51$ V.

Fe^{3+}/Fe^{2+}, $E^o = +0.78$ V.

SAQ 5.2c

Calculate the equilibrium constant for the reaction of metallic zinc with copper sulphate solution.

$E^{o}_{Cu^{2+}/Cu} = +\ 0.337\ V$

$E^{o}_{Zn^{2+}/Zn} = -\ 0.763\ V$

Recall that $0.0591 \times \log K_{eq} = n\{E^{o}_{(1)} - E^{o}_{(2)}\}$.

You can see that as the difference in the bracketed term becomes smaller then K_{eq} becomes smaller. You will remember that small values of K_{eq} had adverse effects on our model for neutralisation reactions. Let us therefore examine the effect of the difference in potentials on redox reactions.

Go back to the calculation carried out for the titration of iron(II) sulphate with cerium(IV) sulphate solution; we are to examine similar cases but with cerium(IV) sulphate replaced by hypothetical oxidants. These hypothetical oxidants have progressively smaller values of E^{o}, hence the difference between the electrode potentials becomes smaller.

Note that in spite of the smaller values of E° the hypothetical oxidants are all capable of oxidising iron(+2).

Π Calculate the potential of the system in which 100 cm^3 of 0.100*M*-iron(+2) is titrated with 0.100*M*-oxidant (*a*, *b* and *c*), for which $E^\circ(a) = 1.3$, $E^\circ(b) = 1.2$, $E^\circ(c) = 1.1$ v respectively (Fig. 5.2b).

Carry out the calculation for the addition of the following volumes:

(*a*) 10, (*b*) 95, (*c*) 99,

(*d*) 101, (*e*) 105, (*f*) 110 cm^3.

As this involves a certain amount of repetitive calculation it is best to use a tabular approach.

In the first part there is an excess of iron(+2) and the calculation is the same for each case (*a*), (*b*) and (*c*). Remember that at this stage we use the E° value for the iron couple.

Volume added/cm^3	10	95	99	101	105	110
Ratio	9	0.05	0.01	100	20	10
Log ratio, L	0.954	−1.30	−2.0	+2	+1.3	1
$(-0.0591)L$	−0.056	+0.077	+0.118	−0.118	−0.077	−0.0591
$E^\circ(\text{Fe}) + (-0.0591)L$	0.72	0.86	0.90			
$E^\circ(a) + (-0.0591)L$				1.18	1.22	1.24
$E^\circ(b) + (-0.0591)L$				1.08	1.12	1.14
$E^\circ(c) + (-0.0591)L$				0.98	1.02	1.04

You will find it instructive to examine these results as a normal titration curve as in Fig. 5.2b. The general form paralles our observations for acid-base titrations.

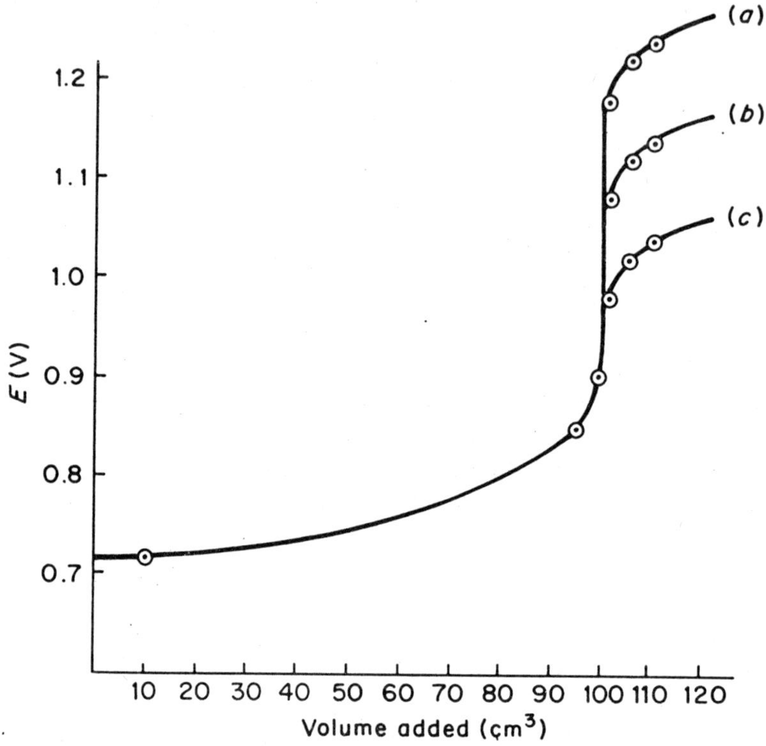

Fig. 5.2b. *Titration of 0.100M-iron(+2) with 0.100M-oxidant*

5.3. MIXED REDOX TITRATIONS

You will recall that in the context of acids and bases we were interested in the variation of pH as neutralisation proceeds. We have also seen that for redox titrations the variation in electrode potential during a titration takes essentially the same form. The potential curve for redox titrations in which the electrode potentials are fairly well separated, by about a volt or more has some similarities with that for the strong acid–strong base titration. The analogy can be pressed a little further by comparing the weak acid–weak base situation with redox titrations in which the difference in electrode potentials is small, say about 0.2 V. Both titrations have no detectable end-point. We shall stay with the analogy as we ask 'What about titrations of mixtures and how do indicators operate?'

In the acid–base titration we noted that the indicators themselves needed to display acid–base behaviour and that in a sense, a neutralisation titration involved two equilibria. The first of these involved the material being determined and the second was, of course, the indicator equilibrium. We saw that if the equilibrium constants were sufficiently different the titration was essentially titration of the analyte and subsequent titration of the indicator. That is we could treat the system as a mixed acid–base system in which the reactions were consecutive. We should expect the same principle to apply to redox titrations. We shall proceed by looking first at redox titration of mixtures without reference to end-point detection, then we shall extend the ideas simply by making one component of the mixture the indicator species.

We now examine a group of *one-electron* reactions to illustrate this. The half-reactions have well separated values of the standard electrode potentials. Note also that we choose to work at $[H^+] = 1.0M$ in this example to simplify the arithmetic.

We consider titration of a mixture of iron(+2) and of titanium(+3) ions by using cerium(+4) as the oxidising agent. The half-reactions are:

$$Ce^{4+} + e \rightleftharpoons Ce^{3+} \qquad E^o = +\ 1.45\ V$$

$$Fe^{3+} + e \rightleftharpoons Fe^{2+} \qquad E^o = +\ 0.78\ V$$

$$TiO^{2+} + 2H^+ + e \rightleftharpoons Ti^{3+} + H_2O \qquad E^o = +\ 0.1\ V$$

What do we expect *qualitatively*?

You can see from the values of the standard electrode potentials that cerium(+4) can oxidise both Fe(+2) and Ti(+3). The question is – does it oxidise them together *or* does it oxidise them in a stepwise manner? The follow-up question (if it turns out to be stepwise) is – in what *order* do the reactions occur?

Recall that the driving force is indicated by the difference in the values for the electrode potentials; consequently we see that the cerium(+4)/Ti(+3) reaction has a higher driving force than the cerium(+4)/iron(+2) reaction. But the question was – does the Ce^{4+} oxidise a bit of both reagents? We can approach this question by assuming that there might be some Fe^{3+} present. If you look at the values of E^o you can see that Fe^{3+} would in fact oxidise Ti^{3+} to TiO^{2+}. Hence as long as there is some Ti^{3+} in solution the iron must remain essentially as Fe^{2+}, ie the reaction is effectively stepwise. This reasoning does, of course, assume that the reactions are quite fast. You might expect that, in general, as the difference in the electrode potentials becomes smaller then the stepwise nature of the reaction(s) becomes increasingly blurred. We noted the same feature when studying stepwise formation of complexes, with the requirement for well separated values of K_{eq}.

The arguments above lead us to the *quantitative* question: are the steps sufficiently distinct to be separately observable?

We now explore this by splitting the reaction into three parts.

(*a*) The region where the principal reaction is the oxidation of titanium(+3).

$$Ce^{4+} + Ti^{3+} + H_2O \rightleftharpoons Ce^{3+} + TiO^{2+} + 2\,H^+$$

(*b*) The region where the principle reaction is the oxidation of iron(+2).

$$Ce^{4+} + Fe^{3+} \rightleftharpoons Ce^{3+} + Fe^{3+}$$

(*c*) The region in which there is an excess of cerium(+4).

The potential at each stage is given by the Nernst equation. Note that in this example we have $[H^+] = 1$, so this term is ignored, as it appears in the denominator of the quotient. Also because we have chosen a one-electron example there are no power terms in the concentrations and hence we can use *ratios* of reactant concentrations and the units of concentration simply cancel.

stage

(*a*) $$E = E^{o}_{Ti} - 0.0591 \log \frac{[Ti^{3+}]}{[TiO^{2+}]}$$

(*b*) $$E = E^{o}_{Fe} - 0.0591 \log \frac{[Fe^{2+}]}{[Fe^{3+}]}$$

(*c*) $$E = E^{o}_{Ce} - 0.0591 \log \frac{[Ce^{3+}]}{[Ce^{4+}]}$$

We shall consider the titration of a mixture of 50 cm^3 of 0.1*M* Fe(+2) and 50 cm^3 of 0.1*M* Ti(+3) with a solution of cerium(+4) sulphate which is also 0.1*M*. The reaction is carried out in decimolar acid solution.

Π Use the same technique as that in the previous example to calculate *E* after the addition of the following volumes: (*a*) 10, (*b*) 40, (*c*) 49, (*d*) 51, (*e*) 75, (*f*) 99, (*g*) 101, (*h*) 110 cm^3.

You will find this fairly easy with a pocket calculator; if you have access to a microcomputer you will find it useful to add a few extra points. Plot the results as a graph.

Your results should look like this.

Stage (*a*) Point	(*a*)	(*b*)	(*c*)
$[Ti^{3+}]/[TiO^{2+}]$ Ratio	40/10	10/40	1/49
Log ratio(L)	0.602	−0.602	−1.69
$(-0.0591)L$	−0.036	+0.036	$+0.099_9$
E	0.064	0.135	0.199_9

Stage (*b*) Point	(*d*)	(*e*)	(*f*)
$[Fe^{2+}]/[Fe^{3+}]$ Ratio	49/1	25/25	1/49
Log ratio(L)	+1.69	0	−1.69
$(-0.0591)L$	−0.099	0	+0.099
E	0.68	0.78	0.88

Stage (*c*) Point	(*g*)	(*h*)
$[Ce^{3+}]/[Ce^{4+}]$ Ratio	100/1	100/10
Log ratio(L)	2	1
$(-0.0591)L$	−0.118	−0.059
E	1.33	1.39

Your results should yield the plot shown in Fig. 5.3a.

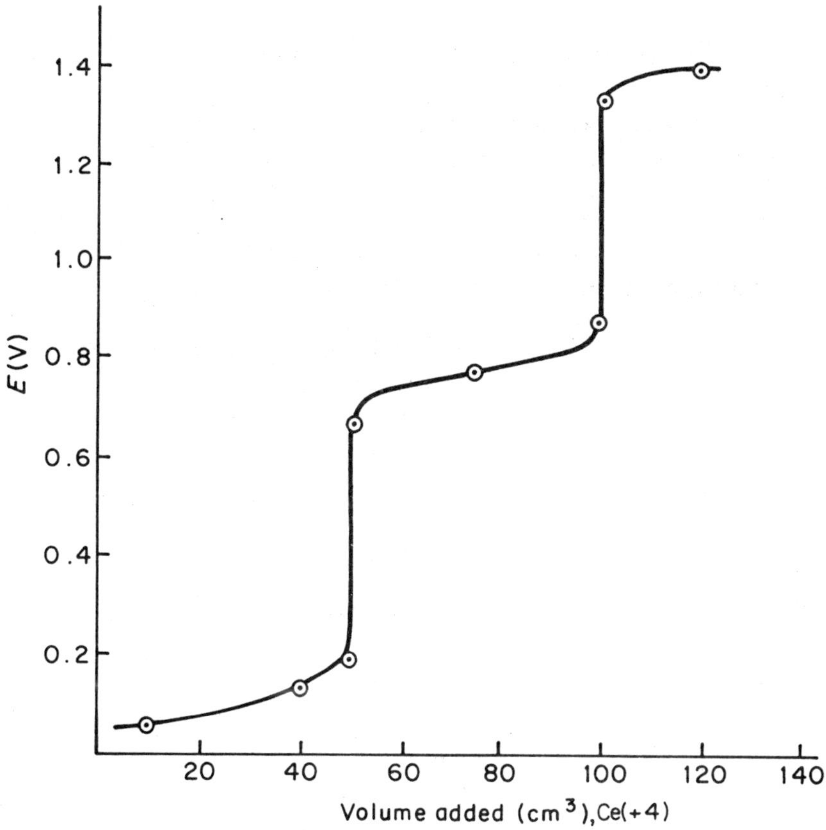

Fig. 5.3a. *Curve for the titration of a mixture of iron(+2) and titanium(+3) with cerium(+4)*

5.4. REDOX INDICATORS

A redox indicator is essentially a compound which has oxidised and reduced forms with different, and preferably intense, colours.

Thus we can represent a general redox indicator reaction by the following familiar half-reaction.

$$\text{Indicator}_{(\text{oxidised form})} + ne = \text{Indicator}_{(\text{reduced form})}.$$

$E^{\circ}_{(\mathrm{ind})}$ is the standard electrode potential, and since for a reversible equilibrium process the Nernst equation is applicable,

$$E = E^{\circ}_{(\mathrm{ind})} - \frac{0.0591}{n} \log \frac{[\text{reduced form In}]}{[\text{oxidised form In}]}$$

we can apply the same reasoning as used in the description of the visual detection of neutralisation indicators. If we say that in general a colour is detectable when it is ten times more intense than other colours which might be present, we see the reduced form when $[\mathrm{In}_{(\mathrm{red})}]/[\mathrm{In}_{(\mathrm{ox})}] > 10$, and the oxidised form when $[\mathrm{In}_{(\mathrm{red})}]/[\mathrm{In}_{(\mathrm{ox})}] < 1/10$. If we now put these values into the Nernst equation we can calculate the approximate range of visual detection of an indicator change in terms of a change in potential,

ie from

$$E = E^{\circ}_{\mathrm{ind}} - \frac{0.0591}{n} \log 10 \text{ to } E^{\circ}_{\mathrm{ind}} - \frac{0.0591}{n} \log (1/10).$$

As $\log 10 = 1$ and $\log (1/10) = -1$ this becomes

$$\text{from } E^{\circ}_{\mathrm{ind}} - \frac{0.0591}{n} \text{ V to } E = E^{\circ}_{\mathrm{ind}} + \frac{0.0591}{n} \text{ V}$$

ie a range of $0.118/n$ V.

The significance of this is that it represents the minimum separation of the 'plateau' parts of the titration curve if end point detection is to be visual. Thus we can see that the selection of a satisfactory indicator for a redox titration does not depend on the exact potential at the equivalence point. Rather it depends on the plateau parts being sufficiently widely spaced for the indicator titration curve to fit neatly between them.

Fig. 5.4a shows (*a*) a titration curve without an indicator present, (*b*) a curve for a large quantity of indicator alone, (*c*) a curve for the mixture. In practice, of course, we use the smallest amount of indicator commensurate with visibility so that the indicator part of the curve reduces to a mere 'blip'.

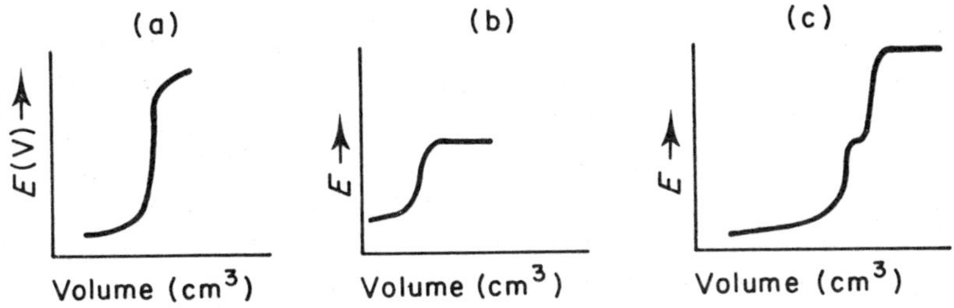

(*a*) No indicator present
(*b*) A large amount of indicator alone
(*c*) Titration curve for the mixture with the indicator

Fig. 5.4a. *Hypothetical redox titration curves, see text*

Fig. 5.4b below lists a few selected redox indicators.

Indicator	E°, pH = 0	Colour change, OX → RED
Indigo monosulphonate	0.26	blue → colourless
Phenosafranine	0.28	red → colourless
Indigo tetrasulphonate	0.36	blue → colourless
Methylene blue	0.53	blue–green → colourless
Diphenylamine	0.76	violet → colourless
Diphenylamine sulphonic acid	0.85	red–purple → colourless
Ferroin	1.06	pale-blue → red
5-Nitro-1,10-phenanthroline-iron(II) complex	1.25	pale-blue → red–violet

Fig. 5.4b. *Some redox indicators*

Selecting the indicator

In practice, for a one-electron change we need a difference of potentials of about 0.4 V, for a two-electron change this is slightly reduced to 0.3 V. Refer to the graph plotted as part of the last exercise (Fig. 5.3a). We can see that the indicator 5-nitro-1,10-phenanthroline-iron(II) complex has a potential which is too high for universal use, although it would be suitable for the titration with cerium. Ferroin would be suitable for (*a*), and might just be acceptable for (*b*), but it would not be appropriate for (*c*).

Objectives

Now that you have completed this Part you will be able to:

- identify the oxidant and reductant in redox processes,
- deduce the change in oxidation numbers during redox processes,
- write balanced equations for redox reactions by using half-reactions,
- relate half-reactions to galvanic cell-reactions,
- use standard reduction potentials to predict chemical changes,
- use the Nernst equation to relate potential and concentrations,
- relate the equilibrium model with the concept of system potential,
- derive titration curves for redox titrations,
- illustrate the relationship between the relevant potentials and the shape of titration curves,
- relate mixed titration curves with indicator behaviour.

6. Volumetric Analysis

6.1. VOLUMETRIC ANALYSIS METHODS

6.1.1. Introduction

Volumetric análysis is an analytical technique of long-standing usefulness. It is a basic component of any practical chemistry course and I am sure you have had experience of both the practical and the theoretical aspect of volumetric analysis. What we shall try to do in the sections that follow is to check that your ideas on technique are sound, and to raise the standard of your knowledge of volumetric analysis to a level worthy of a practising analyst.

In some laboratories volumetric analysis has been downgraded, as have other parts of 'wet chemistry', on the grounds that it is basic, repetitive, time-consuming, and elementary. Many parameters are now measured by instrumental or robotic means, with computer interfacing and control.

Let us examine the arguments that volumetric analysis is inessential today and see why this assertion is false. We hope to show that it has a place in the modern laboratory.

Π What are the uses of volumetric analysis in the industrial laboratory of today?

Examples of present-day usage include the determination of:

— acid content: acid value of oils, acidity in gaseous and liquid effluents, purity of drugs eg aspirin, acids in foodstuffs such as cheese, wines; acids in etching, cleaning, and plating fluids;

— base content: ammonia in fertilisers, base in bleaches, base content of minerals;

— metal content of alloys and minerals, hardness of water;

— redox values: available chlorine and peroxide, traces of oxidants and reductants in foods, unsaturation in oils, vitamin analysis.

— there are, also non-aqueous titrations: Karl Fischer titration for water, phenol content, and lead content.

Π What are the advantages of present-day volumetric analysis?

— It is more precise than many instrumental methods: typically measurements are precise to 0.1% or better.

— The methods are simple, the capital costs are small, and training is easy so that unit costs per estimation are small.

— It works best when one is measuring the major components of a product or mixture. Many instrumental methods measure the minor components of a matrix better.

Π What are its disadvantages?

— As with other systems, considerable time has to be spent in making up standard solutions, standardising secondary solutions, and checking glassware, so that this time expenditure is best justified by long runs of similar samples. Autotitration will shorten the time taken to a slight extent.

— High-quality analytical results are obtained only by good practice and training acquired over a considerable period.

— It is not good for trace components of a mixture: it is better for measuring major components. It is poor in discriminating between similar analytes in a mixture.

Where does that leave volumetric analysis? It is still the method of choice in many laboratories for routine measurements, especially where cost is important.

Even where auto-analysis, use of ion-selective electrodes, potentiometric and other more sophisticated instrumental methods have superseded volumetric analysis, calibration still tends to be done volumetrically. A few examples will help to make the point.

- Unsaturated oils, as found in corn oil, margarine, and other products can be determined spectrophotometrically by ir, uv, and nmr methods but the results are often checked or calibrated against the results from iodometric analysis.
- Chloride in cheese and other products can be estimated by use of an ion-selective electrode, but this method must be calibrated by precipitation reactions.
- Ammonia in fertilisers can be auto-analysed colorimetrically but standards are checked by titration.

I hope these examples have shown that there is a need for volumetric analysis as one of the essential tools any competent analyst should have.

SAQ 6.1a

Consider the example above where the unsaturated-oil component of a commercial vegetable oil is determined either by a uv method or by iodometric titration. Assuming several different unsaturated components are present:

(*i*) What would be the value of an 'iodine number' (measure of the unsaturation) obtained volumetrically?

(*ii*) Why would the uv measurement be less useful?

(*iii*) Suggest a method by which all the unsaturated components in the oil might be determined.

SAQ 6.1a

6.1.2. Method of Titration

Having established that volumetric analysis is an essential component of modern analysis, let us review the basic techniques.

First, what are the characteristics of this type of analysis? There are four main types of reaction used in estimations that can usefully be done by this technique as follows.

- Acid–base reactions, eg ethanoic acid with NaOH;
- Precipitation reactions, eg Ag^+ with Cl^-.
- Redox reactions, eg Fe(II) with MnO_4^-.
- Complexation reactions, eg Ca^{2+} with EDTA.

Each of these could be done in a non-aqueous medium but water is the usual medium. Not every reaction in the above categories lends itself to titrations.

A titration reaction has to be:

(*a*) stoichiometric; proceeding according to a known reaction path and equation(s), with no alternative or side reactions;

(*b*) rapid; ie over in fractions of a second in contrast with say, a colorimetric reaction which takes time to reach equilibrium; also rapid reaction also makes titration less temperature-sensitive;

(*c*) quantitative; i.e 100% complete in the desired direction, or with an equilibrium lying very far to the right as in $Ag^+ + Cl^- \rightleftarrows AgCl$;

(*d*) observable; some property must change so that a sharp change is observable on either side of stoichiometric balance (the *equivalence-point*). A property such as conductance, potential, or pH could be used, or change of colour, either directly or by using the secondary colour-change of an indicator. Especially in this last case the observed *end-point* is not necessarily coincident with the *equivalence-point* because of the delay in getting the indicator to show the change, and other factors. Ideally end-point and equivalence-point should be as close as possible.

Π Are the following reactions in aqueous solution suitable for use in titrations?

(*a*) $CuSO_4 + 2\,NaOH \rightarrow Cu(OH)_2 + Na_2SO_4$

(*b*) $CH_3CO_2H + C_3H_7CO_2Na \rightarrow CH_3CO_2Na + C_3H_7CO_2H$

(*c*) $CuSO_4 + 4\,NH_4OH \rightarrow Cu(NH_3)_4SO_4 + 4\,H_2O$

(*d*) $Na_2CO_3 + 2\,HCl \rightarrow NaCl + H_2O + CO_2$

(*a*) This is unlikely to be suitable because the reaction turns out to be non-stoichiometric with all sorts of basic salts being precipitated.

(*b*) This is an equilibrium reaction with a small equilibrium constant in aqueous solution due to the small pK_a difference, with little change in properties of the solutions, and therefore unsuitable. It might well be better in a non-aqueous solvent such as pyridine.

(*c*) This reaction also involves a series of equilibria but can be used, as the final complex is clearly defined by its colour, so that back titration of excess of NH_4OH is just possible.

(*d*) This is a reaction routinely used in a titration, but depends on removal of all the CO_2 which would tend to remain in solution in equilibrium. This removal is done by efficient heating and stirring.

Let us take a simple titration and follow it through stepwise, eg the estimation of HCl by NaOH standardised by use of potassium hydrogen phthalate.

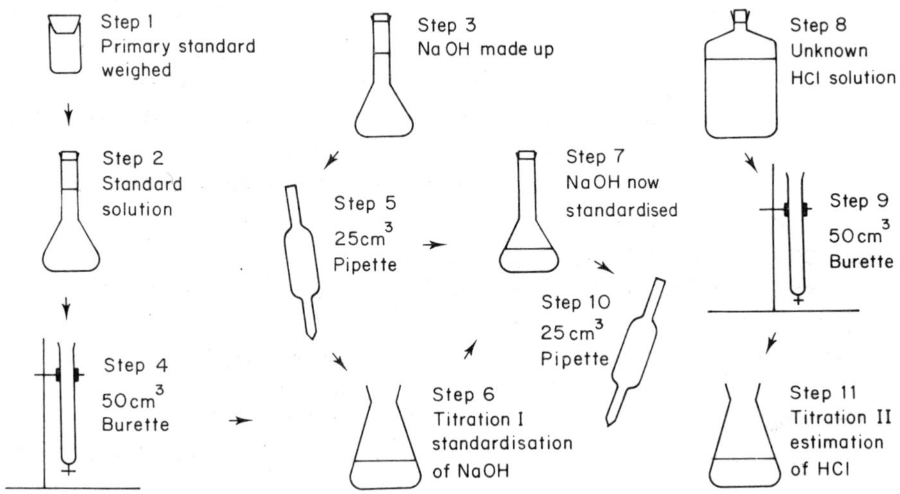

Fig. 6.1a. *Steps in a titration*

Let us assume we have a laboratory with a fairly uniform temperature (preferably 20 °C), clean grade-A glassware, good quality distilled water, and all calibration done. (All these factors are significant and will be discussed later.)

Primary standards

First we have to make up a solution, which is highly accurate in concentration, by using our primary standard.

Available primary standards are:

Potassium hydrogen phthalate (KHP) $C_8H_5O_4K$	M_r = 204.23	Acid/base
Sulphamic acid NH_2SO_3H	M_r = 97.03	Acid/base
Anhydrous sodium carbonate Na_2CO_3	M_r = 105.99	Acid/base
Potassium dichromate $K_2Cr_2O_7$	M_r = 294.19	Redox
Arsenic (III) oxide As_2O_3	M_r = 197.85	Redox
Potassium iodate KIO_3	M_r = 214.00	Redox
Sodium oxalate $Na_2C_2O_4$	M_r = 134.00	Redox
Sodium chloride NaCl	M_r = 58.44	Precipitation
Silver nitrate $AgNO_3$	M_r = 169.87	Precipitation

Primary standard substances are available in a very high state of purity and, after being dried at 105–110 °C, do not either gain or lose mass in a sealed container. It is convenient for weighing if the relative molecular mass (M_r) is high.

Weighing (Step 1 in Fig. 6.1a)

A clean dry weighing bottle half-filled with the primary standard is weighed to 0.1 mg. It is removed from the balance room and an amount close to that required is transferred into a clean funnel inserted in a one-litre volumetric flask. (The use of a rough balance for guidance is acceptable). No transfer by spatula (with possible loss of material) is necessary. The weighing bottle and stopper are reweighed and the transferred mass is calculated.

Π Why not add the solid to an empty weighing bottle?

The main reason is inability to guarantee total transfer from the bottle, as well as adding material to the bottle in the balance room, which is undesirable.

Making up the solution (Step 2)

If the solid is readily soluble in cold water it is washed into the flask with distilled water, the funnel is rinsed well and removed, and the volume made up to near the mark. The stopper is inserted and the flask shaken to dissolve the solid and remove air bubbles. When settled and equilibrated the solution is made up to the mark (lowest part of the meniscus).

If the solid is not so soluble, it is weighed into a 250 cm^3 beaker and dissolved in warm distilled water with stirring. The solution is allowed to cool to room temperature and transferred by using a glass rod and filter funnel, to the volumetric flask, and the solution is made up to the mark as before.

Π Why not heat the flask to aid dissolution?

Heating the flask may damage it but more importantly will alter its calibration. The glass will take a long time to return to ambient temperature and it will take even longer for the volume to return to its original value.

Use of burette (Step 4)

A clean 50 cm^3 grade A burette is set up vertically in a suitable stand with a beaker and white tile beneath. The burette is lowered to a suitable position so that it can be filled easily with the phthalate solution by using a small funnel. The solution is allowed to drain through and the waste is collected in a beaker. It is now refilled to about 2 cm above the zero mark, the funnel is removed, bubbles are removed from below the tap or in the body of the burette and solution is run out to the zero graduation. A burette reader such as that in Fig 6.1b makes this easier. The end of the burette is touched to the beaker to remove excess of solution from the end of the burette.

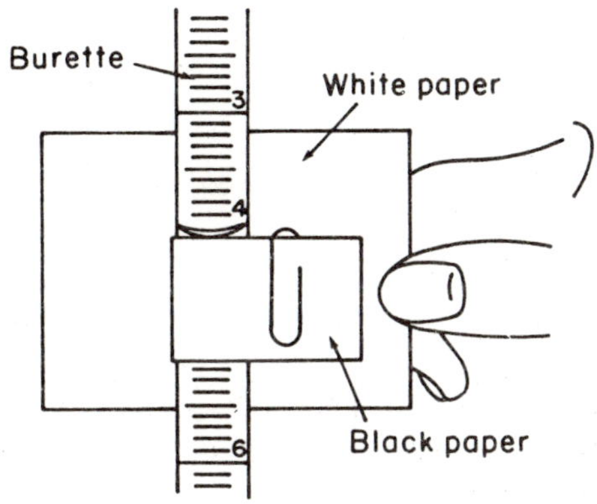

Fig. 6.1b. *Reading a burette*

Use of pipette (Step 5)

A clean dry 25 cm^3 pipette is used. With mouth suction or preferably rubber-bulb suction the sodium hydroxide solution is drawn up from the beaker of solution to just beyond the graduation. This portion is let go to waste. The process is repeated but this time the fluid is released to the mark as in Fig. 6.1c and any excess is wiped away on the side of the beaker.

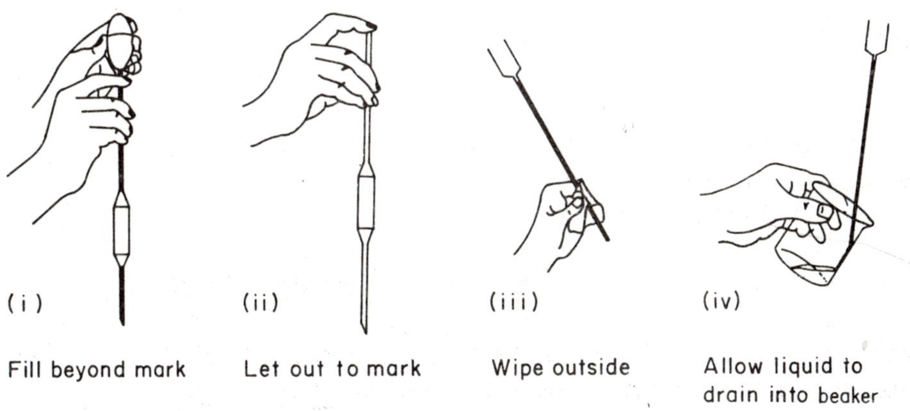

Fig. 6.1c. *Using a pipette*

- Draw liquid past the graduation mark.

- Use forefinger to maintain liquid level above the graduation mark.

- Tilt pipette slightly and wipe away any drops on the outside surface.

- Allow pipette to drain freely.

The residual solution is allowed to run into each conical flask in turn. The last drops of solution must not be blown out, but the pipette touched to the side of the flask.

Π Why not pipette directly from the graduated flask?

Have you tried it?

Air will be sucked up as the pipette won't reach into the body of the flask.

Π Why not blow out the last drops?

- The pipette is calibrated to include the small residue left on touching the inside of the conical flask.

- Blowing into a pipette makes it dirty.

- CO_2 in the breath may react with the solution being pipetted.

Titration (Step 6)

One or two drops of indicator are added and the samples are titrated by using a technique similar to that shown in Fig. 6.1e, with slower addition near the end-point.

Liberal use of distilled water from a wash bottle is advised as well as efficient swirling. The end-point should be obtained by the addition of just one drop of titrant.

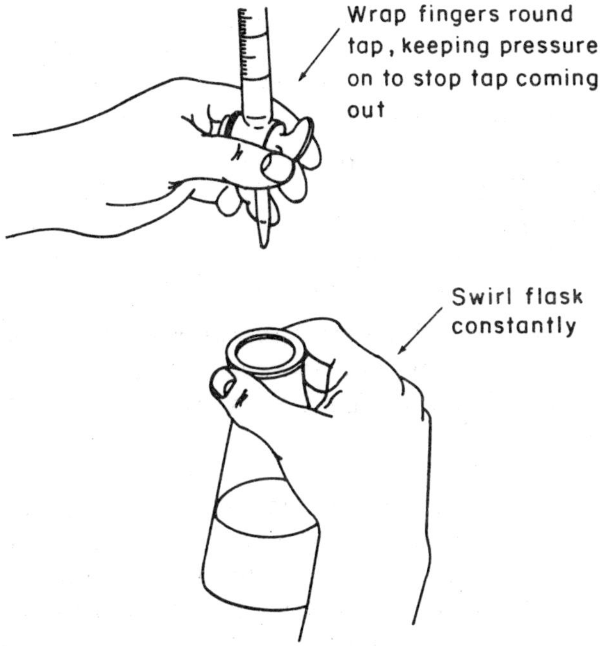

Fig. 6.1d. *Titration Procedure*

The volume used is measured to the nearest 0.01 cm^3 (if possible) and noted. The burette is refilled to the zero mark and the titration flasks are washed out with distilled water.

Many people have difficulty with dexterous swirling during titration (as in Fig. 6.1d)! Only practice will improve it.

In a similar fashion, the result from steps 7 to 11 (Fig. 6.1a) for the estimation of the hydrochloric acid is obtained.

Π Why is sodium hydroxide never used in the burette although ammonium hydroxide might well be?

Caustic soda solution causes sticking of glass joints and mild etching of the glass

The equipment is now drained, washed with water and distilled water, cleaned further if necessary, and allowed to dry at room temperature

6.1.3. Units of Concentration

The mole

For the purposes of volumetric analysis we need to be clear on the units of mass, volume, and molecular amount. All masses are based on the kilogram although masses are conveniently expressed in grams.

As we are studying stoichiometric reactions we should be clear about the mole. 1 mole is defined as that mass in grams containing an Avogadro Number of molecules or atoms. One mole is thus 6.023×10^{23} molecules or atoms. As an example, one mole of sodium chloride ($M_r = 58.44$) contains 6.023×10^{23} molecules of sodium chloride and 1 mole (or gram atom) each of sodium and chlorine (each 6.023×10^{23} atoms).

The molar mass is defined relative to a fixed mass of a standard (now 1/12th of the mass of the pure Carbon-12 isotope). For most volumetric analysis purposes this is the same as the old atomic weight and molecular weight based on hydrogen.

Volume

Of the possible units of volume cm^3, ml, l, dm^3, m^3 which are relevant to us here, cm^3, dm^3 = (1 litre) are either acceptable to SI (Systéme Internationale d'Unités) or, if not are still in current usage; ml is falling out of use.

Π What is the difference between 1 cm^3 and 1 ml?

1 ml = 1.000028 cm^3, the difference arising from the fact that a millilitre is 1/1000th of the volume of 1 kg water at its maximum

density, and cm^3 is based on the standard unit of length, the metre, kept at Sévres. For our purposes we can still use the equivalences $1\ dm^3 = 1000\ cm^3 = 1$ litre.

Concentration

Arising from what we have just said, we are likely to have concentrations in terms of mass per unit volume or moles per unit volume. Referring again to our sodium chloride example ($M_r = 58.44$), 58.44 $g\ l^{-1}$ (58.44 $g\ dm^{-3}$) is the concentration, equivalent for this compound, to 1 mole per litre.

This latter mode of expression is the basis of the *molarity* system, where a $1M$ (molar) solution is defined as one of a concentration of 1 mole of a substance in a total volume of one litre of solution.

Π Does this mean there are 1000 cm^3 of water present?

No. The volume of water often lies between 900 cm^3 and 990 cm^3 depending on the dissolved volume of the solid.

Notice in the molarity system the label $1M$ is always related to a standard one litre of solution. Here are some simple molarity examples to illustrate the point.

Example 1

(*a*) How many moles of NaOH are there in 125.0 cm^3 of $5.00M$-NaOH [M_r(NaOH) = 40.00]?

$5M$ refers to one litre, we want 125.0 cm^3.

$$\frac{125.0}{1000} \times 5.00 = 0.625 \text{ mole}$$

(*b*) What mass of NaOH does this represent?

$$1 \text{ mole} \equiv 40.00 \text{ g NaOH}$$

$\therefore$ 0.625 mole = 40.00 × 0.625 = 25.0 g NaOH.

Example 2

What mass of NaOH is required to make 50.00 cm^3 of 0.1227*M*-NaOH solution?

1 litre of 1*M* solution is made from 1 mole of NaOH (40.00 g in 1 litre) and 0.1227*M* by 0.1227 × 40.00 g in one litre, and therefore a 50.00 cm^3 portion of this latter solution is made from

$$\frac{50.00}{1000} \times 0.1227 \times 40.00 \text{ g NaOH} = 0.2454 \text{ g NaOH}$$

You will have a chance to practise these and other calculations in the next section.

The molal system

The molal system is used in situations where solutions are made up by mass as for industrial acids and bases, as % w/w or w/v, the former being related to molality. Also several physical measurements such as depression of freezing point and other colligative properties require molal solutions

Here, typically, one mole of the compound is mixed with 1000 g of water to make up the solution. As the volume of the solution is unpredictably related to the original volumes, we need extra information, usually the density or specific gravity to work out the molarity. An example or two should help.

Π Commercial ethanoic acid is 99.50% by mass and has a density of 1.051 g cm^{-3} at 20 °C.

(*a*) What is its molality? (M_r = 60.05)

(*b*) What is its molarity?

(*a*) In 100.0 g of ethanoic acid there are 99.50 g acid or 99.50/60.05 moles of ethanoic acid = 1.6569 mole.

These are in 0.500 g of water.

Therefore in 1000 g of water there would be $\frac{1000}{0.500} \times 1.6569$ moles, ie 3314 moles; therefore the solution is 3.314×10^3 molal.

(*b*) Considering 1.000 g of this solution,

mass of ethanoic acid = 0.995 g = 0.01657 mole.

As the density = 1.051 g cm^{-3},

$$\text{volume of } 1.000\ g = \frac{1.000}{1.051} = 0.9515\ cm^3,$$

ie the concentration is 0.01657 mole/0.9515 cm^3.

$$\therefore \quad \text{mass (in moles) per litre} = \frac{1000}{0.9515} \times 0.01657 \text{ moles}$$

$$= 1.74 \times 10^1 \text{ molar}$$

Notice the large values this concentrated acid produces and also the large disparity between the two results. They are semi-independent measures connected only by means of density. Molality is not often used in titrations. In practical situations the concentrated acid above would be diluted (say ×100), estimated, and the molarity of the concentrated acid calculated.

Normality and molarity

A system much in use at one time but now largely superseded is the *Normality* system based on *equivalents*.

If we adjusted M_r by an amount equal to the number of replaceable protons in an acid (or correspondingly for a base) or alternatively the number of electrons associated with a redox reagent in its stoichiometric equation we should get the equivalent for the substance. Thus:

for KOH,1 equivalent = 1 mole;

for $KMnO_4$ ($5e^-$), 1 equivalent = 1/5th mole;

for H_3PO_4 (if $3\,H^+$), 1 equivalent = 1/3rd mole.

Solutions of these gram-equivalents made up to standard volume (1 litre) would give a normal (1*N*) solution

The advantages of this system are first that all reactions occur on an equivalent to equivalent basis and the expression

$$N_1 V_1 = N_2 V_2$$

would simplify calculations of concentration. Also a mixture of fatty acids could be given an overall equivalent value which has some use, while the molarity figure is meaningless for this type of sample.

The disadvantages are, first, that it is an artificial system of fractional molarities, but more fundamentally, each substance does not have a unique equivalent but one dependent on use.

eg H_3PO_4 could yield H^+, $2\,H^+$, or $3\,H^+$, ie 3 different equivalents;

or KHC_2O_4 could yield 1 H^+ ($1M = 1N$), or release 2 electrons per mole in a redox reaction $1M = 2N$.

For these reasons normality systems and equivalents have, to a large extent disappeared. Nevertheless, I have included some normality calculations along with those on molarity and molality in the next section.

6.1.4. Volumetric Calculations

It is little use completing the manipulative side of volumetric analysis successfully if the result, a concentration or mass, is incorrectly worked out. Some people do have difficulty with these calculations, but I hope to show that if we use the mole idea and progress through some examples, the topic is not difficult. You would be well advised to seek other examples for practice to achieve proficiency.

(A) Molarity calculations

Example 1

0.2514 g of a sample containing copper was dissolved in acid, diluted, treated with an excess of potassium iodide and titrated with sodium thiosulphate of concentration 0.1010*M*, 26.32 cm^3 being required. Calculate the percentage of copper present.

Data: A_r for Cu = 63.54 Equations are:

$$2\,Cu^{2+} + 4\,I^- \rightarrow 2\,CuI + I_2$$

$$I_2 + 2\,S_2O_3^{2-} \rightarrow S_4O_6^{2-} + 2\,I^-$$

because

$$Cu^{2+} + e \rightarrow Cu^+, \quad 2\,I^- \rightarrow I_2 + 2e,$$

and

$$2\,S_2O_3^{2-} \rightarrow S_4O_6^{2-} + 2e,$$

each Cu^{2+} reacts to produce iodine equivalent to one $S_2O_3^{2-}$.

Calculation:

26.32 cm^3 of 0.1010*M*-$Na_2S_2O_3$ contain $\frac{26.32 \times 0.1010}{1000}$ moles = 0.002658 mole of thiosulphate.

This equals the number of moles of Cu^{2+} and as A_r(Cu) = 63.54, mass of Cu = 0.002658 × 63.54 g = 0.1689 g, and the percentage of

$$\text{copper} = \frac{0.1689}{0.2514 \times 100} = 67.18\%\ \text{copper.}$$

Example 2

50.00 cm^3 of a sodium carbonate/sodium hydrogen carbonate solution were titrated against 0.1020*M*-hydrochloric acid.

Titration (indicator: phenolphthalein) required 10.00 cm^3 of acid and further titration (indicator: methyl orange) required 25.05 cm^3 of acid. What is the concentration of each component?

Data: With phenolphthalein as indicator,

$$HCl + Na_2CO_3 \rightarrow NaHCO_3 + NaCl$$

Then with methyl orange,

$$HCl + NaHCO_3 \rightarrow NaCl + H_2O + CO_2$$

Calculation: Number of moles of HC1 used in first step = $\frac{10.00}{1000} \times$ 0.1020 = 0.001020 mole.

This will also be the number of moles of Na_2CO_3 in 50.00 cm^3.

$\therefore$ concentration $= \frac{1000}{50.00} \times 0.001020$

= 0.0204M-sodium carbonate.

In the second stage $\frac{25.05}{1000} \times 0.1020 = 0.002555$ mole of HC1 were required.

This is equivalent to that from the original Na_2CO_3 (now $NaHCO_3$ = 0.001020 mole) plus that from the initial $NaHCO_3$ (= 0.002555 − 0.001020 = 0.001535 mole).

This is in 50.00 cm^3. $\therefore$ in 1 litre number of moles $NaHCO_3$ =

$\frac{1000}{50.00} \times 0.001535 = 0.3070$

ie 0.3070M-sodium hydrogen carbonate

or $3.070 \times 10^{-1} M$.

Now try it for yourself.

SAQ 6.1b

25.00 cm^3 of a solution of Ni(II) and Ca(II) were titrated with 0.1025*M*-EDTA at pH5 and required 22.35 cm^3. The pH was then adjusted to 10 and titration required 14.75 cm^3 EDTA. What is the concentration of Ca(II) and of Ni(II)?

Data: A_r(Ni) = 58.71, A_r(Ca) = 40.08

at pH5, Ni^{2+} + EDTA = NiH_2EDTA = $2H^+$

at pH10, Ca^{2+} + EDTA = $CaEDTA^{2-}$ + $4H^+$

SAQ 6.1c

50.00 cm^3 of a Fe(II)/Fe(III) mixed solution were reduced in a Jones reductor to give Fe(II). Before reduction the Fe(II) required 18.34 cm^3 of 0.02000*M*-$KMnO_4$; after reduction 42.34 cm^3 were needed.

Calculate the concentration of Fe(II) and of Fe(III) in the solution.

Data:

$$Fe^{2+} \rightarrow Fe^{3+} + e$$

$$MnO_4^- + 8H^+ + 5e \rightarrow Mn^{2+} + 4H_2O$$

SAQ 6.1d

The chromium in 1.0254 g of an ore was oxidised to $Cr_2O_7^{2-}$ which was treated with 25.00 cm^3 of 0.4000*M*- Fe(II) solution (an excess). This excess of Fe(II) required 34.85 cm^3 of 0.02642*M*-$KMnO_4$. What is the percentage of chromium in the ore?

Data: A_r (Cr) = 52.00

$$Fe^{2+} \rightarrow Fe^{3+} + e$$

$$MnO_4^- + 8H^+ + 5e \rightarrow Mn^{2+} + 4H_2O$$

$$Cr_2O_7^{2-} + 14H^+ + 6e \rightarrow 2Cr^{3+} + 7H_2O$$

(B) Normality calculations

Example 1

30.00 cm^3 of silver nitrate solution were titrated with a solution containing 0.2000 g of potassium cyanide, according to the complexation reaction: $Ag^+ + 2\,CN^- \rightarrow [Ag(CN)_2]^-$. M_r (KCN) = 65.12.

What is the normality of the silver nitrate solution?

Calculation:

$$\text{Number of equivalents of } CN^- = \frac{0.2000}{65.12} = 0.003071.$$

These must react with the same number of equivalents of Ag^+, ie there are 0.003071 equivalents of Ag^+ in 30.00 cm^3

$$\therefore \quad \text{Normality} = \frac{1000}{30.00} \times 0.003071$$

$= 1.024N$-silver nitrate

(Note! molarity is half this).

Example 2

0.6038 g of an iron ore was dissolved in acid and the iron was reduced to the Fe(II) state. Titration required 38.42 cm^3 of 0.1073N-potassium dichromate. Calculate the percentage of iron in the ore.

Data: A_r (Fe) = 55.85.

$Fe^{2+} \rightarrow Fe^{3+} + e$,

$Cr_2O_7^{2-} + 14\,H^+ + 6e \rightarrow 2\,Cr^{3+} + 7\,H_2O$

Calculation: Number of equivalents $Cr_2O_7^{2-} = \dfrac{38.42}{1000} \times 0.1073 =$ 0.0041225.

This is the same as the number of equivalents of iron.

For iron (II), 1 mole $\hat{=}$ 1 equivalent and $A_r = 55.85$,

$\therefore$ number of grams Fe = 0.0041225 × 55.85 = 0.2302 g Fe.

$$\therefore \quad \text{percentage Fe} = \frac{0.2302}{0.6038} \times 100 = 38.13\%$$

SAQ 6.1e

0.2734 g of sodium oxalate was dissolved in water, acidified with H_2SO_4, heated to 70 °C and titrated with potassium permanganate solution (42.68 cm^3). Unfortunately too much was added and the excess of $KMnO_4$ was back titrated with 0.1024*N*-oxalic acid, [ethane-1,2-dioic acid, $(COOH)_2$] and needed 1.46 cm^3. What is the normality of the permanganate solution?

Data:

$5\,C_2O_4^{2-} + 2\,MnO_4^- + 16\,H^+ \rightarrow 2\,Mn^{2+} + 10\,CO_2 + 8\,H_2O$

ie $C_2O_4^{2-} \rightarrow 2\,CO_2 + 2e$

$M_r\ (Na_2C_2O_4) = 134.00$

$MnO_4^- + 8\,H^+ + 5e = Mn^{2+} + 4\,H_2O$

SAQ 6.1e

SAQ 6.1f

0.4671 g of a solid containing sodium hydrogen carbonate was treated with hydrochloric acid (40.72 cm^3). This acid was standardised against 0.1876 g of anhydrous sodium carbonate which needed 37.86 cm^3 of acid.

What is the normality of the acid, and the percentage sodium hydrogen carbonate in the solid?

Data:

equivalent of $NaHCO_3$ = 84.01, of Na_2CO_3 = 52.99

SAQ 6.1g

0.2500 g of pure potassium chloride was mixed with 0.4500 g of impure barium chloride and the mixture was titrated with 0.1000*N*-$AgNO_3$, requiring 72.30 cm^3.

Calculate the percentage purity of the barium chloride.

Data:

Equivalent of KCl = 74.56, of $BaCl_2$ = 104.12.

(C) Molality calculations

Example 1

(*a*) What is the molality of a commercial ammonia solution quoted as 28.0% w/w NH_3 and specific gravity of 0.898 g cm^{-3}?

Data: M_r (NH_3) = 17.03

(*b*) How must the ammonia solution be diluted to give a solution that is approximately 3 Molar?

Calculation:

(*a*) 1.00 g of solution contains 0.280 g of NH_3 = 0.280/17.03 = 0.01644 moles of ammonia.

This is in 1.00 − 0.28 = 0.72 g (or cm^3) of water.

$$\therefore \quad \text{We have } \frac{0.01644 \times 1000}{0.72} = 22.83 \text{ moles/1000 g water.}$$

ie 22.8 Molal

(*b*) 1.000 g of solution contains 0.01644 mole of NH_3 but 1 g is

$$\frac{1.000}{0.898} = 1.1135 \text{ cm}^3,$$

ie a concentration of $\dfrac{0.01644}{1.1135}$

moles per cm^3 or 1000/1.1135 × 0.01644 moles per dm^3 = 14.76*M*-ammonia solution.

We are required to produce approximately 3*M* ammonia.

$\therefore$ Dilute five times.

SAQ 6.1h

We wish to make 2*M*-sulphuric acid from commercial conc. sulphuric acid (94.0% w/w and density 1.831 g cm^{-3}) by dilution. How much concentrated acid is needed per litre?

Data: M_r (H_2SO_4) = 98.1.

SAQ 6.1i

Calculate the molarity and molality of:

(*a*) 69.0% nitric acid, density 1.409 g cm^{-3},

M_r = 63.01;

(*b*) 85.0% phosphoric acid, density 1.689 g cm^{-3},

M_r = 98.00.

SAQ 6.1i

6.2. PRECISION AND ACCURACY IN VOLUMETRIC ANALYSIS

6.2.1. Error Tracing and Identification

The point of any analysis is to obtain a meaningful result; one from which decisions can be made. Apart from ordinary commercial considerations of quality control, it may be that the results have a medical or legal significance.

Suppose we take a single measurement of a given quantity. We would hope that it was *accurate* that is, as close as possible to the true value. With fundamental quantities such as Avogadro's Number we are reasonably confident of its value (and also of the error), but in real life we often have only a vague idea of the true value. One way of finding this value is to perform a large number of replicate determinations. This inevitably gives a spectrum of results, clustered together, we hope, about a mean which should be close to this value. If all is well, this spread of results will be small, that is the precision is high, and our confidence in the final value higher than otherwise. There may be a reason why all our results are consistently in error. This bias may be due to *controllable, constant, determinate* errors or to *variable, random, indeterminate* errors.

A pictorial representation of accuracy and precision may help a little.

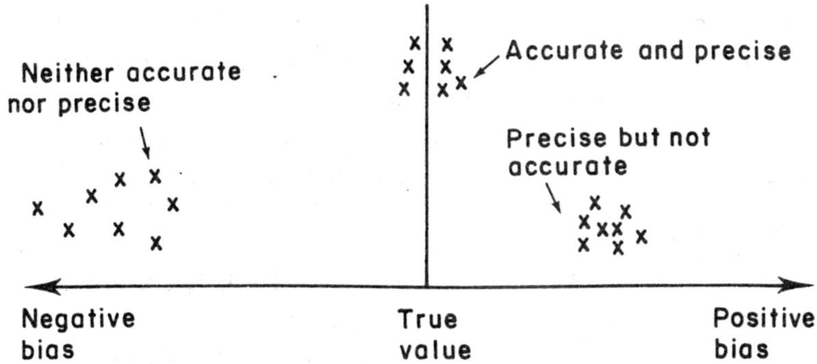

Fig. 6.2a. *Accuracy and precision*

From Fig. 6.2a it can be seen that there may be a net bias or a variable bias as a result of determinate or random errors of various magnitudes interacting, cancelling, or adding in complex ways. In this Section we shall try to identify the causes of error in volumetric analysis and later we shall try to quantify them.

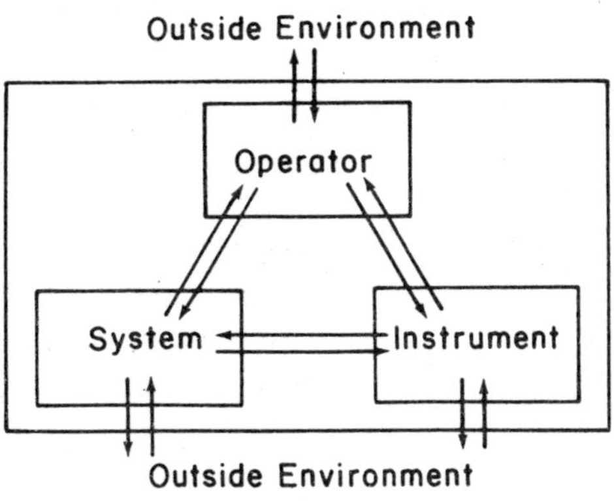

Fig. 6.2b. *The analytical process*

Fig. 6.2b shows an abstraction of the analytical process. The operator is usually human, though an auto-analyser has a similar role, and interacts with the outside environment, the instruments, and possibly the system itself. All other components similarly interact.

To reduce error and be scientific we must control as many of the variables as possible. I have compiled a list of possible interactions causing error in volumetric analysis in Fig. 6.2c. Many of the comments apply to other types of analysis. The list is not exhaustive and you are asked to add to the list yourself.

Sources of error	Action to limit error
Environment	
variable temperature	isothermal lab (preferably 20 °C)
variable pressure	compensate
variable moisture	closed lab and bottles
variable wind	draught-free lab
variable dust	cleanliness
variable light	dark bottles
oxygen	sealed bottles
vibration	insulation
The System, eg solids and solutions	
purity	high purity
water quality	high quality
lack of homogeneity	good mixing
activity effects	activity coefficients
corrosion, decomposition, evaporation	sealed flasks
indicator	blank value
Instruments (Glassware, balances)	
dirt	cleanliness
balance faults	balances checked
glassware faults	glassware calibrated
Operator	
poor weighing poor end-point level misread dirty apparatus grade B equipment	correct practice and technique (training)

Fig. 6.2c. *Sources of error*

Π Apart from the items mentioned already what steps would you take to reduce error to a minimum?

Your answer may include several of the following and others:

- check results against known standards,
- calibrate equipment (see later),
- check reagents and measure impurity levels (eg use Analar and Aristar grades from BDH Ltd),
- measure water quality,
- ensure isothermal conditions,
- check operator by regular 'blind' standard samples, cross measurements, and analysis after standard addition, check balances.

Π How can we classify errors?

Errors can be:

- constant (determinate) errors: impurities in the water, balance bias, large volume of indicator (use blank);
- proportional errors: impurity in a standard, interfering ions, effect of temperature on volume; or
- random errors due to a multiplicity of causes in the environment, instrument, or operator.

As we approach the limit of measurement in a given situation, for example weighing less than 10 mg on an ordinary balance, we reach the region of relatively large indeterminate errors. We have to choose a method appropriate to the precision of result required, and consider factors of cost and time. Thus there is little point reading a net weighing to five or six significant figures if we are going to titrate (to 3 or 4 significant figures). None the less we should aim as high as possible with measurement standards. An excellent though humbling experience is to cross check one method against another for the same analyte. I say humbling, because the results are often different (at an appropriate level) due to the differing errors inherent in each method. So it is common practice in industry to adopt an agreed technique, apparatus, and method.

SAQ 6.2a

Suggest methods for comparison with the determination of dilute hydrochloric acid by acid–base titration.

We now examine some of the methods used to reduce errors, and learn how errors may be calculated and quantified.

6.2.2. Volumetric Glassware

Before going on to the calibration of glassware it might be a good idea to have a closer look at the glassware involved in volumetric analysis.

Glassware

In Britain graduated flasks, pipettes, and burettes are made according to British Standard specifications in two classes; Grade A and Grade B, Grade A being superior to Grade B.

Typical tolerances are shown in Fig. 6.2d

Graduated flasks	25.00 = $\pm$ 0.06 cm^3
	250.00 = $\pm$ 0.30 cm^3
	1000.00 = $\pm$ 0.80 cm^3
Burettes	10.00 = $\pm$ 0.02 cm^3
	50.00 = $\pm$ 0.06 cm^3
	100.00 = $\pm$ 0.10 cm^3
Pipettes	10.00 = $\pm$ 0.04 cm^3
	25.00 = $\pm$ 0.06 cm^3
	100.00 = $\pm$ 0.12 cm^3

Fig. 6.2d. *Typical tolerances of Grade A equipment*

Normally we use a 1 litre graduated flask, 50 cm^3 burettes and 25 cm^3 pipettes.

SAQ 6.2b

> **Suppose we made up a solution in a one-litre flask, pipetted 25 cm^3 out of it and titrated it with 25 cm^3 of some other substance from a 50 cm^3 burette.**
>
> **What is the maximum error due to the tolerances given in Fig. 6.2d.**

In an effort to improve these tolerances various newer designs of burette and pipette are available.

Pipettes

The mechanical pipettes shown in Fig. 6.2e have their own tolerances as specified by the individual manufacturer and most work on a piston-plunger principle. The accuracy and precision of delivery is either quoted by the manufacturer or can be measured by weight calibration. The chief advantage is one of convenience. Safety would be improved when bromine or other hazardous solutions are being handled.

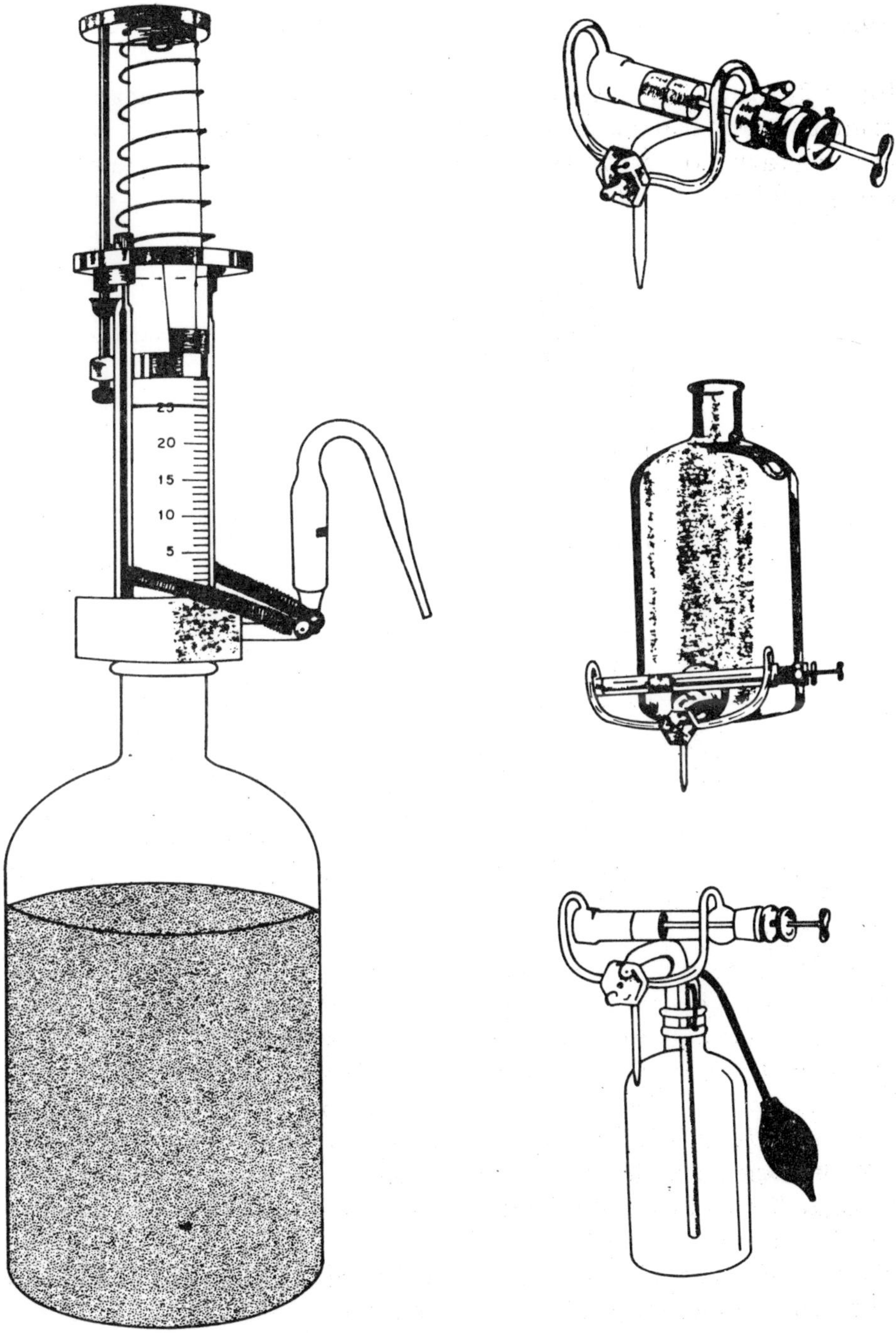

Fig. 6.2e. *Mechanical pipettes*

Burettes

Fig. 6.2f shows variants on the standard burette, once again used principally for convenience in a busy laboratory. Here, by a suitable use of gravity or pressure and operation of taps the burettes are easily refilled.

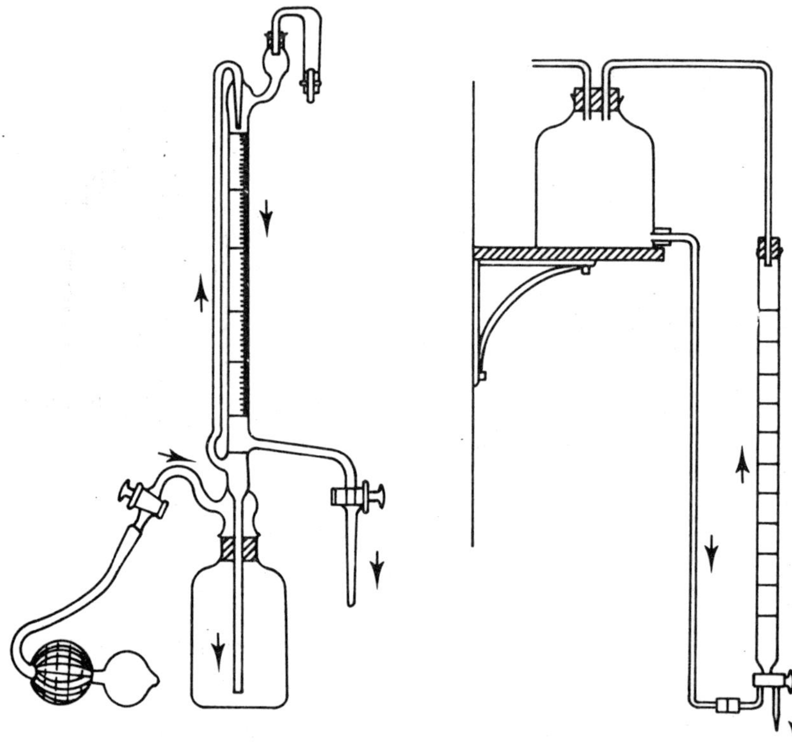

Fig. 6.2f. *Refillable burettes*

Alternatives to the mechanical pipette are weight burettes such as those in Fig. 6.2g. As the name suggests they are capable of being weighed on a balance (useful when molal solutions are required). Volatile liquids such as ethanol solutions, or hazardous liquids such as bromine may be handled in this way.

Fig. 6.2g. *Weight burettes*

Π Is there any problem in using a non-aqueous solvent such as ethanol in a burette?

I can think of two concerns. First, does it attack the tap lubricant? Secondly, with a different surface tension the meniscus and volume delivered will be different.

Cleaning of Glassware

Various solutions are suggested for the cleaning of glassware. In good usage there is unlikely to be any film which is not easily removed. The only ones requiring special treatment are silver nitrate stains removable with dilute aqueous ammonia and permanganate stains removable with conc. hydrochloric acid then water. Mixtures involving bromine or nitric acid, especially with organic solvents, are to be avoided as they can cause explosions.

3% Teepol solutions and other commercial preparations can be used. If more aggressive cleaning is needed, immersion in concentrated chromic acid can be used. It must be remembered that chromic acid, made from $Na_2Cr_2O_7$ and conc. sulphuric acid, is toxic and highly corrosive.

After use glassware should be drained, washed with water and distilled water and allowed to dry at room temperature. If hot-air drying is used the apparatus *must be allowed to equilibrate* with laboratory temperatures before use.

Finally two things to avoid. One is the use of an excess of tap lubricant (not used with Teflon taps) and the other is the use of strong caustic solutions which tend to cause glass joints to seize up.

Water

Often we neglect the quality of water, a major component in volumetric analysis. It is little use having high quality solids if the water is not satisfactory.

Water should be free of particulates, and have as low an ionic and organic content as possible. The total dissolved solid measurement, on evaporation of water, should give a solid content less than 1 mg l^{-1}, but the method is tedious so that the more convenient conductivity values are used. Satisfactorily pure water has a conductivity less than 10^{-6} ohm^{-1} cm^{-1} at 25 °C. The organic content is harder to measure but stability to $M/2000$ acidified potassium permanganate is a good guide. It is a good deal harder to remove species derived from glass such as sodium ions and soluble silica. To do this we would use ion exchange columns and plastic containers.

For most purposes filtered water, distilled or doubly distilled, with or without ion exchange suffices.

6.2.3. Calibration of Glassware

We have already mentioned manufacturer's tolerances for glassware. We must now add to that the effect of the environment on glassware and then set about calibrating it and allowing for both these factors. The factors are significant. Whenever we weigh a flask in air we get a very slight buoyancy effect (usually 1 in 10,000 or less). Of greater importance (possibly 1 part in 1,000) is the effect of temperature on glassware and on the solutions in the glassware (the greatest effect may be 1 part in 500).

We have already decided that a laboratory should be isothermal to limit variations of this sort and that a standard and comfortable temperature is 20 °C. We have to ask ourselves if the making up of solutions and use of burettes particularly will alter if the temperature is 17 °C. As volumes stated on glassware are quoted for 20 °C we must work out the volumes at 17 °C. Usually these factors are worked out, on first use of the equipment and are assumed to be constant thereafter (unless very accurate work is called for). If volatile solvents are used temperature will show yet another effect.

Let's have a look at the two expansion effects. A flask full of solution at 15 °C is allowed to warm up to 19 °C. In doing so the glass will expand slightly (depending on the type of glass) and the liquid level would, for this reason alone, fall. But the aqueous solution will expand by a greater amount so that a net expansion is noted. Our normal concentration (for 15 °C will be wrong and a proportional error will result on pipetting out the solution (NB a similar but smaller effect will arise with the burette). Fig. 6.2h shows the net effect of the two expansions for two types of common laboratory glass.

Temperature	Soda glass	Borosilicate glass
5 °C	998.63	998.39
10 °C	998.76	998.60
15 °C	999.30	999.16
20 °C	1000.00	1000.00
25 °C	1001.30	1001.11
30 °C	1002.31	1002.46

Fig 6.2h. *Volume (cm^3) of 1000 cm^3 water in a flask calibrated for 20 °C*

To interpret this let's look at an example.

The liquid in a flask labelled 1000.00 cm^3 at 20 °C is made up to the graduation at 15 °C. According to the table its actual volume is 999.30 cm^3 and so concentrations calculated from it will be greater by 7 parts per 1000. As we have said earlier this is of the order of the precision of the weighing but an order less than that of titration. Combining these effects with the manufacturer's tolerance effects means we need to check the true volume of a piece of glassware at the nominal operating temperature of the laboratory. Once done this correction can be printed on the flask for subsequent use.

Here we will use Fig. 6.2i and Fig. 6.2j.

Temp (°C)	Mass* (g)	Volume of 1 g of water cm^3
10	998.39	1.0016
11	998.32	1.0017
12	998.23	1.0018
13	998.14	1.0018
14	998.04	1.0019
16	997.80	1.0022
18	997.51	1.0025
20	997.18	1.0028
22	996.80	1.0032
23	996.38	1.0034
24	996.38	1.0036
25	996.17	1.0038
26	995.93	1.0041
28	995.44	1.0046
30	994.91	1.0051
32	994.35	1.0057
34	993.75	1.0063

* Mass of pure water to fill flask.

Fig. 6.2i. *Flask (labelled 1000.00 cm^3) of soda glass, coefficient of cubical expansion 0.000025 (°C)*

Temp (°C)	Mass* (g)	Volume of 1 g of water cm^3
16	997.86	1.0021
18	997.54	1.0025
20	997.18	1.0028
22	996.78	1.0032
24	996.33	1.0037
26	995.85	1.0042
28	995.32	1.0047
30	994.76	1.0053
32	994.17	1.0059

* Mass of pure water to fill flask.

Fig 6.2j. *Volume occupied by 1.0000 g of water in borosilicate glass, coefficient of cubical expansion 0.000010 °C*

As you can see the mass of water required in a 1 litre flask, so that the water will be 1000.00 cm^3 in volume at 20 °C is quoted.

Examples should help again.

Example 1. A graduated flask of soda glass is filled with water to the mark at 16 °C: the contents weighed 998.13 g. What is its volume at 16 °C and at 20 °C?

The mass to give one litre at 16 °C = 997.80 g from the table.

$$\therefore \quad \text{volume at 16 °C of 998.13 g} = \frac{998.13}{997.80} = 1000.33 \text{ cm}^3$$

ie an error of +0.33 cm^3

$$\text{If warmed to 20 °C the volume is} = \frac{998.13}{997.18} = 1000.95 \text{ cm}^3.$$

ie an error of +0.95 cm^3

We could label the flasks accordingly.

Example 2. A borosilicate burette, from the zero to the 10 cm^3 mark delivers a volume of water weighing 9.9878 g. What is its volume correction factor?

From the table:

at 16 °C the mass of a nominal 10.00 cm^3 (quoted at 20 °C)

= 9.9786 g

$$\therefore \quad \text{The volume is } 10 \times \frac{9.9878}{9.9786} = 10.0092 \text{ cm}^3$$

ie a correction of 0.0092 cm^3

We could then measure different 10 cm^3 portions from the burette and produce calibrations. As the error in the above is about 0.01 cm^3, our limit of observation in titration, it may not be worth using. The exercise is necessary though to discover this.

SAQ 6.2c

In a typical laboratory the contents (water) of a borosilicate 100 cm^3 pipette were weighed and found to be 99.588 g at 24 °C. What is the error in volume?

6.2.4. Quantification of Error in Volumetric Analysis

Before we proceed with error determination it would be well to summarise the section on error tracing. We found there that errors were possible from the operator, from the glassware and balances, from the laboratory atmosphere and surroundings, and from the chemicals and solutions used, or from any combination of these.

Put yourself in the role of laboratory manager. You will have standardised the method and procedures and checked the equipment regularly. You will need to check the competence and performance of the laboratory staff and keep an overall picture of the general analytical performance of the laboratory.

First some trivial points. The results should be presented according to the accepted house style to the required degree of precision. (Precision here means the number of significant figures). In this respect, care must to be taken over significant figures, for example 0.001 ± 0.020 is acceptable in some circumstances (the method error of 0.02 is greater than the observed result 0.001), especially when generated by a calculator or when logarithms are used.

Types of error As discussed previously these can be determinate or indeterminate, proportional or constant. At regular intervals standard samples should be submitted for analysis, preferably reflecting the range of samples likely to be met.

Accepted (true) values	Technician A	Technician B
5.26	5.29 (+0.03)	5.28 (+0.02)
11.74	11.76 (+0.02)	11.82 (+0.08)
26.43	26.32 (−0.11)	27.06 (+0.63)
40.57	40.51 (−0.06)	42.11 (+1.54)

Fig. 6.2k. *Comparison of results from two technicians*

A glance at the results shows that technician A appears accurate within about 5 parts per 10000 and the spread of results is probably a random one with results above and below specification. Efforts could be made to check techniques for all types of error. With computerisation of results routine checking of this sort is facilitated. Technician B's results are within about 3% which may be within tolerance, but there appears to be a constant proportional error which needs investigation, a faulty burette being a possibility.

On a single result cross-checked once, we can quote an absolute error or a relative error (in per cent, parts per thousand, etc). Both are important because, for example, where the analyte is present in small amounts the relative error may be high but quality control may require tolerance related only to absolute error.

Standard deviation

So far we have dealt only with accuracy, but when multiple readings are possible and convenient, as for titrations, we can apply statistics to give a measure of the precision but *not* of the error.

Theoretically, if we had a large number of replicate analyses it would give us a truer picture of the mean value and precision of readings then with less. With enough readings we would approach a Gaussian or *Normal Distribution* as in Fig. 6.21.

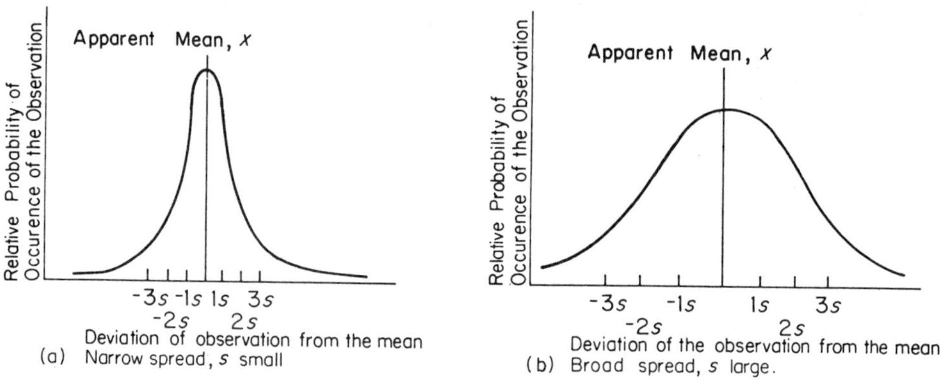

Fig. 6.21. *Normal Distribution of readings*

The precision is measured by the spread of results, here indicated by values of s, the estimated *standard deviation*. It is a mathematical property of such a curve that 68% of all readings should lie between $+s$ and $-s$, and 96% between $+2s$ and $-2s$. The apparent mean is $\bar{x}$ in Fig. 6.21, and is an estimate of the true value μ.

Let us use a simple example to calculate $\bar{x}$ and s. Suppose titration results were 0.2041, 0.2049, 0.2039, 0.2043.

The mean $\bar{x}$ is easily calculated as

$$\frac{0.2041 + 0.2049 + 0.2039 + 0.2043}{4} = 0.2043$$

The standard deviation, s, is calculated from a formula such as

$$s = \left(\frac{\Sigma(x - \bar{x})^2}{(n - 1)}\right)^{\frac{1}{2}}$$

Here n, the number of readings, is 4 and $\bar{x}$ is 0.2043.

x	$x - \bar{x}$	$(x - \bar{x})^2$
0.2041	−0.0002	4×10^{-8}
0.2049	+0.0006	36×10^{-8}
0.2039	−0.0004	16×10^{-8}
0.2043	0	0

The sum of squared deviations $\Sigma(x - \bar{x})$

$$= 56 \times 10^{-8}$$

$$\frac{\Sigma(x - \bar{x})}{(n - 1)} = \frac{56 \times 10^{-8}}{3}$$

$$= 18.67 \times 10^{-8}$$

$\therefore$ Standard deviation, s, $= (18.67 \times 10^{-8})^{\frac{1}{2}}$

$$= 4.3 \times 10^{-4},$$

ie $s = 0.0004$

If s is small the results are clustered and apparently precise, if it is very large it is a cause for concern and may well be a function of the method used.

Π Is the mean result $\bar{x}$ reliable?

As we accumulate replicate values close to the mean we become more confident of it, but always remember that there may be a hidden constant error, ie $\bar{x}$ and μ differ.

Is s a good measure of error?

Yes it is a reasonable measure of random error, which is more reliable with more results.

t and *Q* tests

When we have all our results tabulated we need to ask some fundamental questions about their validity.

The rogue result (*Q* test)

We all have had the situation where one titration reading seems odd and we have quietly forgotten about it. This is not really valid. It might be that the effect of this odd reading is to bring $\bar{x}$ closer to the correct value, μ. The Q test is a suitable means of deciding whether to cast out a result or not for small sample runs. If we had a good idea from a lot of replicate results what the mean and standard deviation values were, then we would have grounds for judging the result; ie the error in a result within $\pm 2s$ of the mean is 95% likely to be random, and the result is valid.

In the Q test the ratio of the isolation of a result ie (the result minus the nearest other value) to the overall range for a given number of measurements, is compared with a calculated Q^{90} value from tables. The Q^{90} value is one for 90% confidence. If the result from the above is less than the quoted Q^{90} value, we judge that the result should be included to 90% confidence.

For example:

For the readings 0.2041 0.2049 0.2039 0.2043 the mean, $\bar{x}$ is 0.2043 and s = 0.0004. If we wish to examine the 0.2049 result by the Q^{90} test we do it as follows:

Range = 0.2039 − 0.2049 = 0.0010
Isolation = 0.2049 − 0.2043 = 0.0006
and the ratio = 0.6.

The Q^{90} value for 4 observations is 0.76 from tables and as this is the greater the result 0.2049 should be retained.

Comparing two sets of results (the paired *t* test)

Suppose two technicians perform the same determination and produce two means and standard deviations for the one sample. Unless we have definite external evidence that one result is superior, for example, results from a standard institution, both are equally believable. What we have to ask is whether they are compatible and whether the spread is due to random factors rather than determinate error.

Let's use an example where:

technician A got $\bar{x}$ = 0.2043 and s_A = ±0.0004, 4 readings;

technician B got $\bar{x}$ = 0.2037 and s_B = ±0.0005, 4 readings.

Then we work out two expressions:

First $(\bar{x}_A - \bar{x}_B)$ which here is 0.0006,

and secondly

$$t^{90}\left(\frac{s_A{}^2}{n_A} + \frac{s_B{}^2}{n_B}\right)^{\frac{1}{2}}$$

From tables t^{90} for 4 + 4 readings (less 2 degrees of freedom) at the 90% confidence level = 1.943.

$$\left(\frac{s_A{}^2}{n_A} + \frac{s_B{}^2}{n_B}\right)^{\frac{1}{2}} = \left(\frac{0.0004^2}{4} + \frac{0.0005^2}{4}\right)^{\frac{1}{2}} = 0.00032$$

$\therefore \quad t^{90} \times$ this value = 0.00062 (the t^{90} factor).

If the difference in means $(\bar{x}_A - \bar{x}_B)$ (0.0006) is less than the t^{90} factor (= 0.00062), which it is (just), then we can say the differences between the two sets are due to random effects alone to 90% confidence.

Often if we compare results from two different methods for the same analyte by this test we find they are not compatible because of different determinate and indeterminate errors. We then have to adopt one as standard and apply corrections to the other result.

A more detailed discussion of the statistical processing of results is given in the Unit of ACOL: Measurement Statistics and Computation.

Corrections

When it is established conclusively that a constant error is involved it is valid to apply a correction factor. The commonest example in volumetric analysis is the indicator correction. Most indicators take a finite volume of reagent to change colour. This amount is found by a blank titration with indicator and titrant alone and the factor is then determined. For example in silver nitrate titrations a correction of about 0.1 cm^3 is common when potassium chromate is the indicator.

Control charts

A chief chemist will want to monitor the quality of analysis of all his instruments and technicians. This is done by means of a control chart which is a simple visual indicator of day-to-day performance.

The results for a standard are measured regularly along with other samples and the results are plotted daily. The accepted (true) value (or agreed value if not known) is shown as well as its standard deviation when measured according to the prescribed method. An example of a control chart is shown diagramatically in Fig. 6.2m.

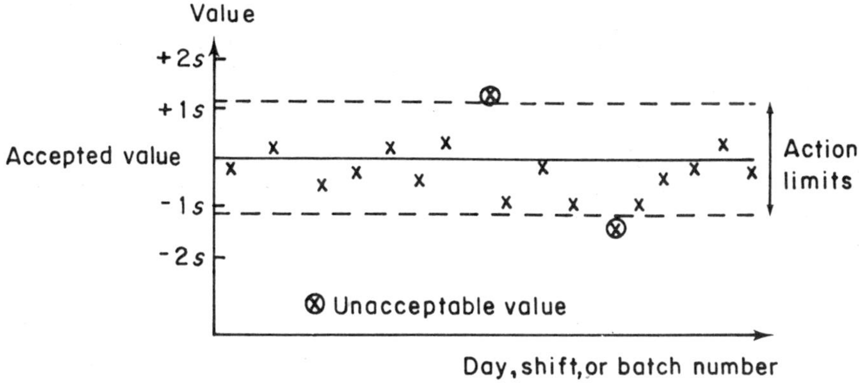

Fig. 6.2m. *A control chart*

The control chart can be used in two ways.

If samples give values outside agreed action limits ($\pm 1.2s$ in Fig. 6.2m) the samples may be sent for retest or the measurement queried. Computerised results can be 'flagged' for human attention. Error tracing would then come into play to trace the cause. Perhaps of more concern is when there is a steady drift in the pattern of errors which suggests something is going off specification, for example a progressively faulty balance, and error tracing would be activated. In volumetric analysis the former situation is more likely.

The statistical analysis of analytical results is a sophisticated process which we have only touched on briefly.

Summary

We have studied in this part the basic elements of volumetric analysis and the need for careful operations to ensure a meaningful practical result. Although the techniques are familiar to most of us, the necessary skills for high grade work are acquired only after a few years experience and constant auditing, and with continual improvement of skill levels.

Features of volumetric analysis that usually need improvement include apparatus calibration, which is often ignored, and routine titration technique, which often deteriorates because of its familiarity. Regular laboratory audits are a good idea to keep skills suitably honed.

The processing of analytical results can be a sophisticated operation crucial to any reliable analysis scheme. We have given only a basic treatment in this part.

Finally, titration calculations are a seperate skill which once mastered presents no trouble. It goes almost without saying, that practice makes perfect, and more titration calculations should be sought and attempted.

7. Acid–Base Titrations

Introduction

In this Part we shall try to put together the theoretical knowledge on acids and bases, gained earlier in Part 2 and the practical aspects of volumetric analysis such as calibration, standardisation and the correct usage of glassware as discussed in Part 6. This should enable you to meet all the likely demands in industrial and laboratory situations for acid–base titrations. Consequently before you proceed further it might be a good idea to look back to the above parts to refresh your memory, although many of the ideas will be mentioned again here.

Nowadays equipment in most laboratories has moved on from the simple burette-pipette-indicator type of operation to automated, semi-automated, or computerised titrations with an interfaced pH or potentiometric facility, so that a specimen can be automatically sampled, titrated with the changing pH or potential being measured and the equivalence-point being determined electronically. The result, after suitable processing, is displayed graphically or on a VDU or even sent to control some parameter in a chemical process as a feed-back signal.

It is nevertheless, still of value to examine the basic ideas of acid–base titrations to avoid fundamental errors. The correct choice of solvent, titrant, indicator and equivalence-point are crucial and many mistakes are possible. In particular, it is important not to take the readout from an autotitrator or computerised system without a careful check on the soundness of the theoretical basis, the calibration of the instrument, and the correct operation of all components.

7.1. PRIMARY STANDARDS, PREPARATION AND STORAGE

As we have already discussed in Part 6, Primary Standards for acid–base titrations have to satisfy several criteria.

(*a*) Purity.

The substance should be obtainable in a pure form and be readily and reliably purifiable.

(*b*) Stability.

The compound should suffer no decomposition at temperatures below 100 °C, nor should there be any weight gain or loss in air, due, for example to water loss or gain or reaction with CO_2 in air.

(*c*) Impurity levels.

These should be specified to an accuracy which allows compensation for their presence if necessary. The impurities should be uniform from sample to sample and obviously be at as low a level as possible. The impurities should be less than 0.01% w/w in total.

(*d*) High Equivalent.

So that reasonable amounts are to be weighed, the relative atomic or molecular mass (or equivalent mass) of the compound should be as high as possible.

(*e*) Physical Properties.

The standard should preferably be a solid for ease of handling, and be readily soluble in the solvent normally used.

(*f*) Reaction rate.

The relevant reaction of the primary standard must be stoichiometric and rapid, so that complete reaction is assured.

Sodium carbonate (anhydrous) Na_2CO_3	$M_r = 105.99$
Sodium tetraborate (anhydrous) $Na_2B_4O_7$	$M_r = 201.28$
Potassium hydrogen phthalate $C_8H_5O_4K$	$M_r = 204.23$
Benzoic acid $C_7H_6O_2$	$M_r = 122.12$
Hydrochloric acid (constant bp) HCl	$M_r = 36.465$

Fig. 7.1a. *Primary standards for acid–base titrations*

An ancillary secondary standard sometimes mentioned is hydrated sodium tetraborate, $Na_2B_4O_7.10\,H_2O$, $M_r = 381.44$.

Π On which criteria do

(*a*) Na_2CO_3 (anhydrous)
(*b*) benzoic acid,

fail as primary standards according to the list given.

(*a*) You should comment on the ability of sodium carbonate to absorb water, as it is a drying agent on its low relative molecular mass and the inconvenience of drying it at 100 °C.

(*b*) Benzoic acid fits most of the criteria but its solubility in water is low, although in non-aqueous solvents such as ethanoic acid (acetic acid) or ethanol this is not so.

Π Apart from benzoic acid just mentioned which other standard from the list is ideal for most non-aqueous solvents?

The answer is potassium hydrogen phthalate which has a high M_r and sufficient solubility in these solvents.

To preserve primary standards in a pristine condition, storage in sealed glass containers is advisable, and where moisture causes problems, in a desiccator.

7.2. pH MEASUREMENT

In Part 2 some of the basic pH theory was outlined in terms of equilibria, and it would be a good idea for you to revise these ideas, which are presented again in outline below, before we move on to discussing the practical aspects of pH measurement.

7.2.1. Basic pH Theory

pH is defined by the Sörensen equation

$$\text{pH} = \log \frac{1}{[\text{H}^+]} = -\log [\text{H}^+]$$

The negative sign is for the convenience of representing concentrations below molar (with which we usually are dealing) in terms of a limited positive scale. H^+ as used above is understood to stand for whatever hydrated forms of the acidic proton exist in solution. Normally, the main form is thought to be the hydroxonium ion (H_3O^+).

We could also have a pOH scale where

$$\text{pOH} = \log \frac{1}{[\text{OH}^-]} = -\log [\text{OH}^-].$$

$[H^+]$ and $[OH^-]$ are connected by the relationship

$$[H^+][OH^-] = K_w.$$

K_w, the ionic product for water, is a constant at a given temperature, and has a value close to 1×10^{-14} at 25 °C. We can see that

$$pH + pOH = 14 \quad \text{at } 25\ °C$$

and so only one scale is needed, for acidity and alkalinity as shown in Fig. 7.2a. Note that for concentrations of acid greater than or equal to 1 M, pH has a zero or negative value.

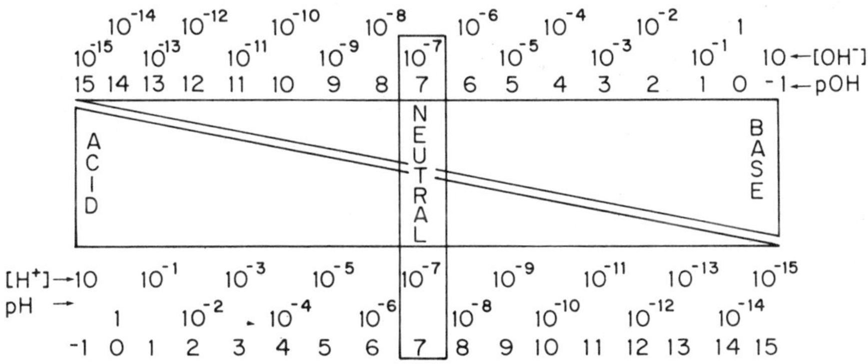

Fig. 7.2a. *The pH and the pOH scale*

In practice solutions of concentrated acids or alkalis are rarely fully ionised and activity factors come into play. More correctly we should relate pH to the activity of H^+ but in dilute solutions the correction is small

$$\text{True pH} = \log a_{H^+} = pH_{obs} + 0.004$$

Also, we should note in passing, that potentiometric measurements involve a_{H^+} rather than $[H^+]$.

Π Why do potentiometric readings relate to a_{H^+}?

The answer is that potentials depend on the nature of the ion as it exists in solution with hydration shells or whatever so that activity controls the readings.

When we mention H^+ we are usually referring to H_3O^+, but in non-aqueous solvents the solvated acidic proton will be different; for example, NH_4^+ in liquid ammonia and $CH_3CO_2H_2^+$ in ethanoic acid, as we shall see later.

I think we should check whether you are happy with pH calculations so some are offered below for you to check your facility with them.

SAQ 7.2a

What is the pH of a solution which is

(*a*) 4.32×10^{-4} *M* in H^+
(*b*) 5.12×10^{-2} *M* in OH^-?

SAQ 7.2b

> **If pH = 8.32 what is**
>
> **(*a*) the H^+ concentration,**
> **(*b*) the OH^- concentration?**

7.2.2. Standard pH

As we have just said, a_{H^+} can be determined potentiometrically with a suitable cell system. A system we could use is one comprising a hydrogen electrode and reference electrode, *viz*.

H_2,1 atmosphere pressure | Pt : H^+ solution | salt bridge | reference electrode

We must realise that the value of pH and $[H^+]$ so measured and calculated is a function of the cell used and the activities of the participating solutions, and of the liquid interfaces, and has no absolute value. We must therefore agree a standard and relate all pH measurements to this closely defined standard.

In Britain a solution of 0.5000M potassium hydrogen phthalate(KHP) is defined as having a pH of 4.000 at 15 °C when we use the following cell.

$$1 \text{ atm } H_2 : Pt \,|\, 0.05000\ M\text{-KHP} \,|\, 3.5\ M\text{-KCl} \,|\, \text{reference electrode}$$

The solution to be measured against this standard uses the cell:

$$1 \text{ atm } H_2 : Pt \,|\, \text{unknown } [H^+] \,|\, 3.5\ M\ \text{KCl} \,|\, \text{reference electrode}$$

ie the same hydrogen pressure, salt bridge, and reference electrode. This gives:

$$\text{pH unknown} - 4.000 = \frac{E \text{ unknown} - \text{E standard}}{0.0591}$$

where RT/nF of the Nernst expression

$$= \frac{0.0591}{n} \text{ at } 25\ °\text{C} \quad (n = 1)$$

The standard cell EMF has a temperature coefficient defined by pH

$$4.000 + \frac{1}{2}\frac{(t - 15)}{100}, \quad t \text{ in °C}$$

Π Is this temperature coefficient the same as the variation of pH with temperature?

No. This is the variation of the standard cell EMF with temperature, which includes many factors such as the Nernst-equation temperature-factor, whereas with pH-temperature variation, it is mainly the change of K_w with temperature.

In America one standard is defined by the National Bureau of Standards (NBS) also based on potassium hydrogen phthalate, but others are defined at various temperatures as in Fig. 7.2b.

Primary Standards	pH at 0 °C	10 °C	20 °C	30 °C	40 °C	50 °C
K H tartrate (saturated at 25 °C)				3.55	3.55	3.55
0.05 molal K H Phthalate	4.00	4.00	4.00	4.02	4.03	4.06
0.025 molal KH_2PO_4 /0.025 molal Na_2HPO_4	6.98	6.92	6.88	6.85	6.84	6.83
0.01 molal $Na_2B_4O_7.10\,H_2O$	9.46	9.33	9.23	9.14	9.07	9.01
Secondary Standards						
0.05 molal K tetroxalate	1.67	1.67	1.68	1.69	1.70	1.71
Calcium hydroxide (saturated at 25 °C)	13.43	13.00	12.63	12.30	11.99	11.70

Fig. 7.2b. *National Bureau of Standards agreed pH solutions*

The solutions quoted in Fig. 7.2b are made up on a molal basis, ie so much mass of substance per kilogram of water.

Π Why are they made up on a molal basis?

The answer is most likely to be convenience: allowances for expansion of the solution and container do not have to be made. Hence in Fig. 7.2b the concentrations would be written 0.05 *m* rather than 0.05 *M*.

7.2.3. pH and Temperature

pH, as discussed above, is defined at a suitable temperature and corrections are based on the Nernst equation or empirical formulae for actual cells. These corrections are measurement corrections, due to the variation of electrode potential with temperature. Thus the voltage-response range of a cell system with a fixed standard and measuring solution will vary with temperature.

More fundamentally, we have to realise that the pH range say between 0.1 M H^+ and 0.1 M OH^- is not always 12 pH units (pH1 to pH13). This, as stated earlier arises from the fact the K_w, the ionic product for water, like most equilibrium constants, varies with temperature. This variation is shown in Fig. 7.2c below.

Temperature °C	0	10	20	30	40	50
$10^{14} K_w$ (moles^2 dm^{-6})	0.12	0.29	0.68	1.47	2.92	5.47

Fig. 7.2c. *The variation of the values of* K_w *with temperature*

For a dilute solution a_{H^+} remains reasonably constant but pH, pOH, and neutral pH vary with temperature so that for example, at blood temperature (37 °C) neutrality is pH 6.8,

$$\text{ie } -\log [H^+] = -\log [OH^-] = 6.8 \text{ not } 7.0.$$

This means that when we make a pH measurement at a temperature other than 25 °C we should first make an 'instrumental' correction for temperature, then measure the pH, and then correct this to a pH at standard temperature (25 °C)

Π After calibration of a pH meter for temperature a solution was found to have a pH of 6.9 at 40 °C. Is it acidic or alkaline?

From Fig. 7.2c, at 40 °C, $K_w = 2.92 \times 10^{-14}$, ie p$K_w$ = 13.535.

so for neutrality ($[H^+] = [OH^-]$),

$$pH = 13.535/2 = 6.77$$

and our solution of pH 6.9 is thus basic relative to the neutral pH of 6.77.

7.2.4. pH Meters

A typical pH meter is shown in Fig. 7.2d below.

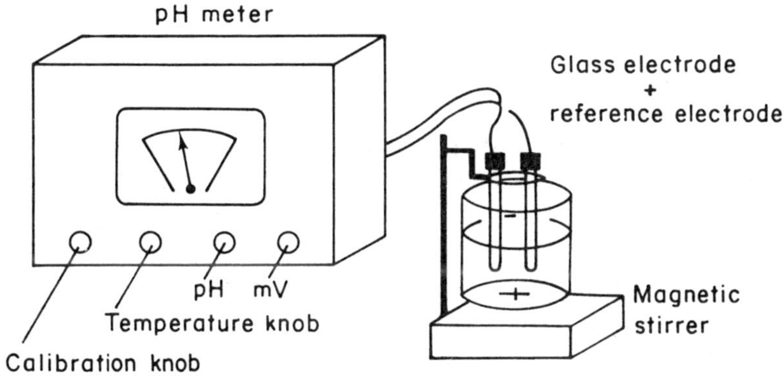

Fig. 7.2d. *pH meter*

Here a glass pH-electrode and a reference electrode are placed in the solution the pH of which is to be measured and the potential difference after equilibration (often the solution is stirred magnetically) is recorded on a low-current meter, often directly calibrated in pH units. In everyday use we check the instrument against buffer solutions of known pH, and adjustments are made for any system errors such as 'alkali error', 'acid error', and salt effects which we shall come to shortly.

We now go on to consider possible electrode systems for measuring pH and study the glass electrode in some detail, as it is by far the commonest system in use today. What are the essentials of a pH-meter system? First we need an *indicator electrode* sensitive to the concentration or more properly the activity of the H^+ ion, then a *reference electrode*, insensitive to the measured solution, good electrical connection between the electrodes, and a suitable low-current device for measuring the potential developed.

Π Why do you think a low-current detector is built into most pH meters?

The answer is because the electrode system presents a high resistance so that for a fixed potential only low currents will be produced. Low currents also mean that any electrolytic effects will be low.

7.2.5. Electrode Systems

The ultimate standard half cell for most cell systems is also a possible indicator electrode for H^+. This is the Standard Hydrogen Electrode. The potential of this is by definition, zero at 25 °C, if hydrogen gas at 1 atmosphere pressure is in contact with H^+ at an effective concentration of 1.000 *M* (a_{H^+} = 1.000)

Fig. 7.2e is a schematic diagram of the Hydrogen Electrode.

It has several advantages amongst which are

- it gives an absolute measurement;
- it works across the whole pH scale without alkali, acid, or salt errors;
- it does not need a liquid junction, with its associated problems (we shall discuss these later)

At the same time it is

- inconvenient and bulky,
- sensitive to oxidising agents such as oxygen,
- sensitive to catalyst poisoning and many metal ions.

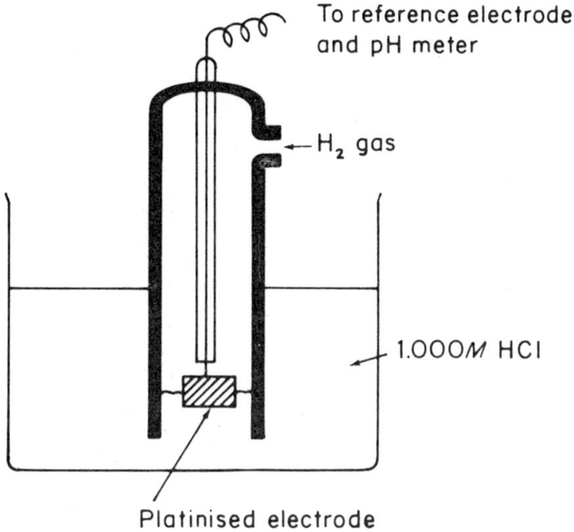

Fig. 7.2e. *The hydrogen electrode*

The half cell equation for the hydrogen electrode is

$$E = E^{o} - \frac{2.303\ RT}{nF} \log \frac{(\text{pressure } H_2)}{a_{H^+}}$$

But since E^o at 25 °C is defined as zero and $\frac{2.303 \text{ RT}}{\text{nF}}$ at 25 °C = 0.0591 (n = 1),

$$E = -0.0591 \log \frac{1}{a_{H^+}} = -0.0591 \text{ pH}$$

Several models of hydrogen electrodes are available and the Lindsey type is illustrated below in Fig. 7.2f.

The apparatus is first flushed with hydrogen and the solution to be measured is poured in *via* 1, so that the level of the liquid is just above the platinum electrode. The salt bridge (2) is added and when a measurement is made a little hydrogen gas is let into the system *via* the tap to cause a slight oscillation of liquid level. Any excess of hydrogen can escape *via* the gas trap at 3.

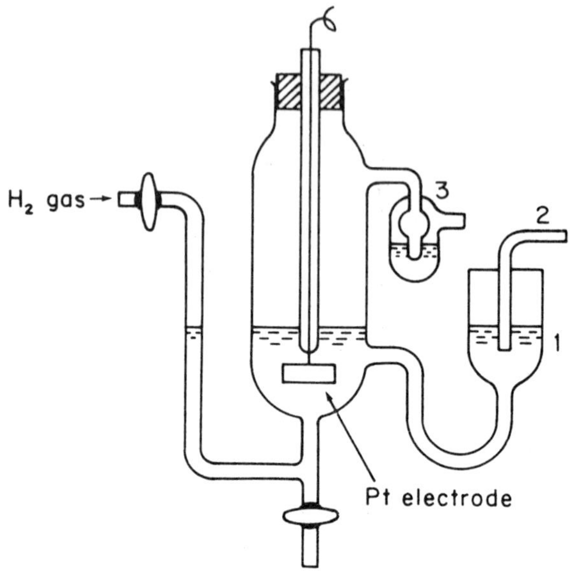

Fig. 7.2f. *The Lindsey hydrogen electrode*

Π Why is this oscillation in liquid level required?

The main reason is to help equilibration of hydrogen gas and acid at the metal surface. The reading settles down more quickly. Note that the apparatus has to be drained and cleaned before a second sample can be measured.

In theory we could link two such hydrogen electrodes, one having standard conditions and the other being the indicator electrode. This would be in effect a concentration cell. Unfortunately the potential generated might be masked by the junction potential of the salt bridge. What exactly is this liquid-junction potential?

Bridges are necessary to join two different solutions electrically when it is not possible or desirable to allow the solutions to mix. The connection is usually made by using a concentrated salt solution kept in place by various means. An asbestos wick soaked in a salt solution, porous glass, capillary connections, or most commonly a fritted glass U-tube may be used. The commonest salt used is potassium chloride (in concentrated aqueous solution), alone or mixed with a setting agent such as agar.

The potential of the liquid junction arises out of unequal charge distribution at the ends of the bridge (junctions). Ideally a bridge allows the free flow of any ion equally in either direction, but in practice discriminates principally on the grounds of ionic mobility. H^+ and OH^- are fast-moving ions relative to most others including K^+, Na^+, Cl^-, SO_4^{2-} and as we get unequal flow of ions a potential will build up at each of the two junctions (the potentials do not cancel out). This junction potential can be as much as 40 mV which may be equal to or greater than the measured potential. If we use a concentrated solution of potassium chloride (eg 3.5 *M* or saturated) the junction potentials are much lower (*ca* 10 mV) as K^+ and Cl^- have approximately equal mobilities.

Π Will the H^+ and OH^- still cause a potential at the junctions?

They tend not to, as most ionic transport is by the K^+ and Cl^- ions, which are usually in great excess relative to the H^+ and OH^- ions.

However we must always bear in mind that the junction potential E_J is a variable function of pH, ionic strength of the solution, and temperature, as all these affect ionic mobilities.

The hydrogen electrode as described above is both a reference electrode and an indicator electrode. However, its use is limited as it is clumsy and subject to interferences as we mentioned before.

Possible alternative systems for hydrogen ion measurement include those in Fig. 7.2g.

	Indicator electrode	Reference electrode
No-junction cells	Hydrogen	Silver* or Calomel
Liquid junction cells	Antimony	Silver* or Calomel
	Quinhydrone	Silver* or Calomel
	Glass	Silver* or Calomel

* Also called the silver–silver chloride electrode.

Fig. 7.2g. *Electrode systems for hydrogen ion measurement*

The Junctionless Cell

For this system we might use, as indicated in Fig. 7.2g a Hydrogen Electrode as indicator electrode with a reference silver–silver Chloride half-cell dipping into the same solution. The complete cell is specified below.

$$\text{Pt}: \text{H}_2 \text{ gas} \mid \text{H}^+ \mid \text{AgCl(solid)} \mid \text{Ag}$$

At 25 °C,

$$E_{\text{cell}} = E^{\text{o}}_{\text{AgCl/Ag}} - E^{\text{o}}_{\text{H}^+/\text{H}_2} - 0.0591 \frac{a_{\text{H}^+} . a_{\text{Cl}^-}}{\{\text{p(Hydrogen gas)}\}^{\frac{1}{2}}}$$

Π Would there be a difference between measuring the pH of hydrochloric acid as a sample, and that of any other acid?

Yes there would. The overall potential depends on the $[H^+]$ (or a_{H^+}) and the $[Cl^-]$ or (a_{Cl^-}), also on the pressure of hydrogen which would be kept approximately constant at 1 atmosphere. We should expect a greater variation when using HCl as both a_{H^+} and a_{Cl^-} rather than just a_{H^+} are variable.

The system is not very convenient with 3 possible variables; there is however no junction potential to complicate matters.

Cells with Liquid Junctions

If the previous junctionless cell was split into two compartments connected *via* a salt bridge, we should then have a cell with a junction and hydrochloric acid could be used in the hydrogen half as few chloride ions would pass across the junction to affect the silver–silver chloride electrode. Potassium nitrate or sulphate could be used in the salt bridge. If it contained KCl and were found to leak, this would affect the potential of the silver–silver chloride electrode, but not if it contained KNO_3 or K_2SO_4.

Hydrogen Indicator Cells

As listed in Fig. 7.2g we have several alternatives to the hydrogen electrode as potential hydrogen-indicator electrodes. Let's look at these in turn, and appraise their practicality for $[H^+]$ (or a_{H^+}) measurement.

The Quinhydrone Electrode

Here we use a couple of organic compounds which interchange in a redox reaction involving H^+.

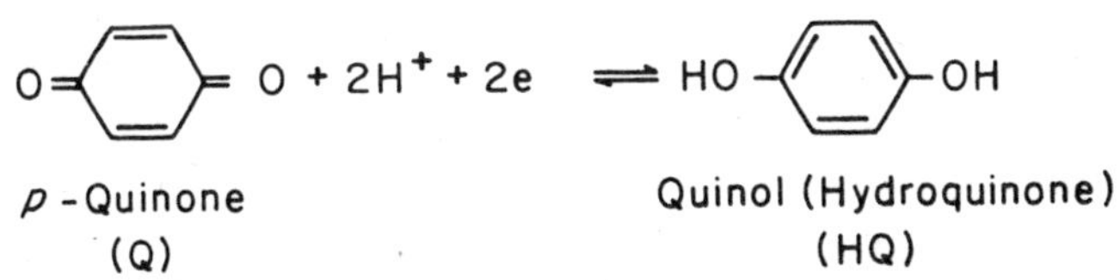

Quinhydrone is a 1 : 1 mixture of the two.

The practical set-up is to have a platinum electrode dipping into an unknown solution to which an equimolar mixture of Q and HQ (quinhydrone) is added, and the electrode is connected via a salt bridge to a reference electrode. The standard half-cell potential at 25 °C is 0.700 V so that the half-cell equation is: $E = 0.700 + 0.0591 \log a_{H^+}$. The solution will be rather messy and of no further use and must contain no other analytes such as chlorine, which might affect either substance. In alkaline solution the hydroquinone forms salts so this cell is not used above pH 9.

The Antimony Electrode

This hydrogen-ion-sensitive electrode is simply made by exposing an antimony rod to air to give a coating of oxide (mainly Sb_2O_3). This oxide-coated rod when inserted into the test solution, which is connected to a reference cell *via* a salt bridge, acts as an indicator electrode for hydrogen ions.

$$Sb_2O_3 + 6H^+ + 6e \rightarrow 2Sb + 3H_2O$$

$$(E = 0.145 - 0.0591\ \text{pH at } 25\ ^\circ\text{C})$$

The system is remarkably robust so that, only the rod and salt bridge need dip into the sample which might be, for example, a waste-water stream. Disadvantages include its reaction with aqueous solutions of low pH causing it to dissolve, and its sensitivity to redox reagents, especially oxygen.

The Glass Electrode

This is by far the commonest H^+ ion-indicator electrode and a diagram of a typical arrangement is shown in Fig. 7.2h.

It consists of a glass tube with a very thin bulbous end containing hydrochloric acid, which is inserted into the test solution and connected by the usual salt bridge to a reference electrode. The potential developed on the inside of the thin glass wall is different from that on the outside surface and a special type of junction potential is set up. The wet glass surfaces are thought to be partly alkali silicate, and partly hydrated silica (or silicic acid). Alkali metal ions and H^+ ions can be exchanged, resulting in a potential dependent on the $[H^+]$ externally (which is variable) and the $[H^+]$ internally (which is fixed). There is some ion migration through the glass itself.

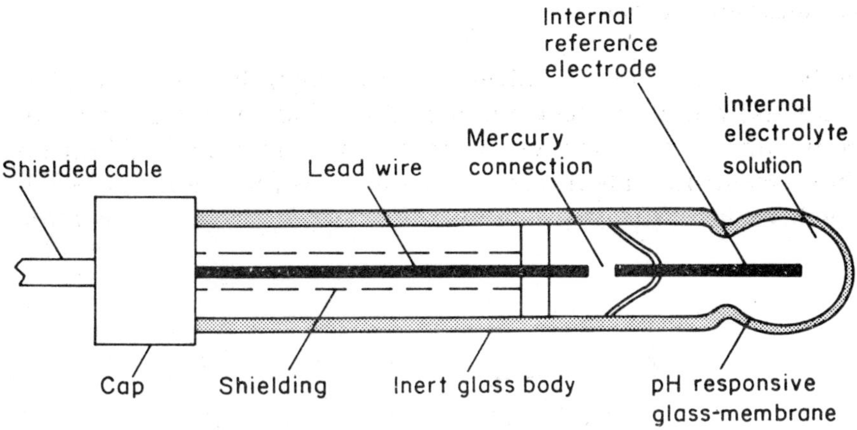

Fig 7.2h. *Typical glass electrode*

The system can be represented as follows.

H^+(internal) | glass | H^+(unknown) | salt bridge | reference electrode

We have, of course, introduced an extra junction with its own characteristics, some of which are similar to the liquid salt-bridge junction. Because the inner and the outer glass surface are bound to generate different potentials (even if the solutions are the same) we get a junction potential which is asymmetric and depends on the type of glass, age, and usage of the electrode. Standard use requires the probe to be left in distilled water before and after use.

$$E = k + 0.0591 \text{ pH}$$

The value of k will thus have to be determined by using a standard solution of known pH or alternatively eliminated as indicated below.

$$\text{pH} = \text{pH(standard)} + \frac{[\text{E} - \text{E(standard)}]}{0.0591} \text{ at } 25\ ^\circ\text{C}$$

Glass Electrode Errors

As the boundary process involves H^+–Na^+ interchange (if the glass is soda glass) the glass electrode is both an indicator electrode for H^+ ions and an ion-selective electrode for Na^+ ions. This means that in concentrated alkali metal ion solutions (eg NaOH or NaCl) there will be an error called the *alkaline error* which is illustrated in Fig. 7.2i.

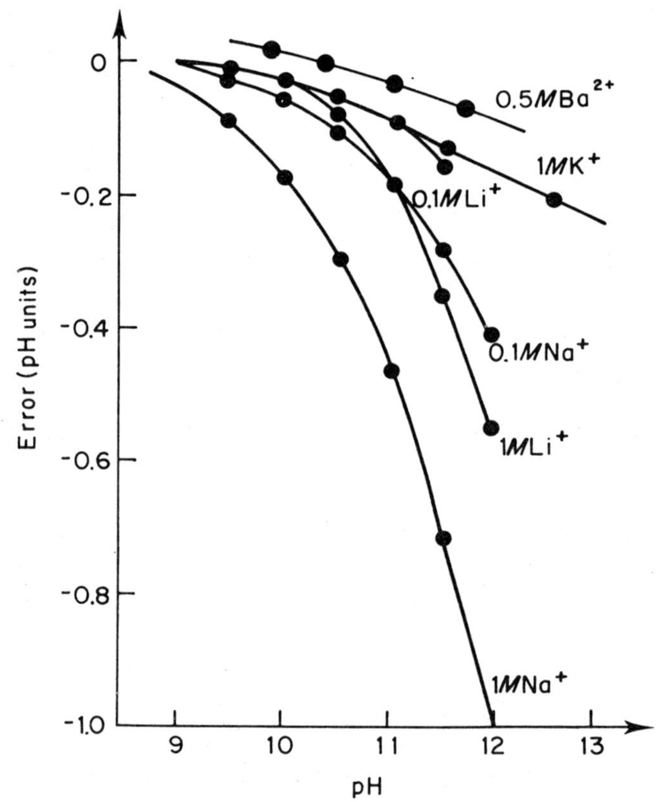

Fig. 7.2i. *Error of the Corning 015 glass electrode in strongly alkaline solutions containing various cations*

Π What will be the apparent pH of a solution of pH 12.0 which is also

(*a*) 1 *M* in Na^+,
(*b*) 1 *M* in K^+,
(*c*) 1 *M* in Li^+,

By using Figure 7.2i you should estimate the pH values to be

(*a*) 12 − 1.0 = 11.0
(*b*) 12 − 0.2 = 11.8 (or just over)
(*c*) 12 − 0.55 = 11.45 (approximately).

An electrode of soda glass with a typical composition of 22% Na_2O, 6% CaO, 72% SiO_2, has a substantial alkaline error, but special glasses, such as that with the composition Li_2O 28%, Cs_2O 2%, BaO 4%, La_2O_3 3%, SiO_2 63%, have been defined for which the alkaline error is very low. In highly acidic solutions an *acid error* may have to be corrected as in Fig 7.2j. This arises principally from activity effects in concentrated solutions.

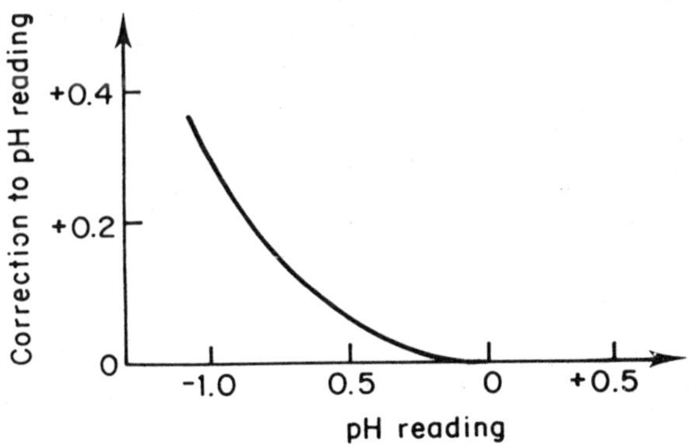

Fig. 7.2j. *Error of the glass electrode in hydrochloric acid solution*

Π What will the pH reading be in 10 *M* HCl?

10 *M* HCl has a nominal pH of −1 (log 1/10) and with the correction of approximately +0.3 from Fig. 7.2j this gives a pH reading of −1 + 0.3 = −0.7. The error is positive rather than negative: a negative error would give a value of say, −1.3.

Despite these errors the glass electrode is the most widely used system today.

Π Will the glass electrode be

(*a*) robust,
(*b*) wide ranging in application,
(*c*) of wide industrial use?

(*a*) It is in fact quite robust,though it can be easily broken. It is often provided with a plastic protector which allows access by the solution.

(*b*) The normal pH range of the glass electrode is pH 1–12 or wider if acid and alkaline errors are corrected, although it works poorly about pH 7 if no buffers are present.

(*c*) It is used widely; it is especially useful for coloured and mixed industrial samples, and is relatively unaffected by redox reagents and metal ions. No wonder it is much used.

The glass electrode does, however, need to have the *k* factor measured regularly. It can also be incorporated into a single probe assembly into which the reference electrode is built to make a combined electrode.

We now briefly examine reference electrodes, the other component used with hydrogen-indicator electrodes.

Reference Electrodes

Apart from the Standard Hydrogen Electrode, two standard reference electrodes are in common use, the calomel electrode and the silver electrode. The calomel electrode is illustrated in Fig. 7.2k below.

The half cell formulation is $Pt \mid Hg \mid Hg_2Cl_2 \mid Cl^-$, and has an electrode potential of:

+0.336 V with 0.1 *M* KCl,
+0.281 V with 1 *M* KCl,
+0.244 V with saturated KCl (35 g/100 g H_2O),

all at 25 °C.

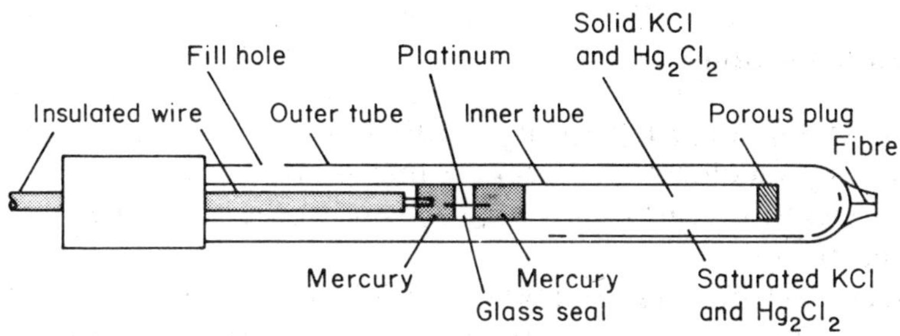

Fig. 7.2k. *A commercial version of a saturated calomel electrode*

Mercury is in contact with mercury(I) chloride paste which is in contact with a KCl solution of known concentration.

Π In the calomel electrode the mercury(I) chloride is shaken with a little mercury and KCl solution before use and the KCl solution is shaken with mercury and a little Hg_2Cl_2 before making it up. Why is this?

For the calomel electrode it is important that cell equilibrium has been fully set up, so all components are shaken together to equilibrate them.

A silver electrode is less elaborate and consists of a silver rod electrolytically coated with chloride and dipping into a KCl solution of variable concentration. The standard potential of this half-cell is 0.290 V if 0.1 *M*-KCl is used, and 0.199 V if saturated KCl is used, (both at 25 °C).

7.2.6. Use of the pH Meter

A typical pH meter was shown in Fig. 7.2d and we now summarise our recently acquired knowledge to arrive at a satisfactory operating procedure.

Operational Steps

- Glass electrode is kept in distilled water.
- Reference electrode checked and ready.
- All solutions allowed to equilibrate to the same measured temperature.
- The electrodes are immersed in a buffer solution of known pH.
- Calibration for temperature response is done, usually automatically by using the temperature knob on the instrument.
- The pH reading is set by using the buffer solutions.
- The probe is rinsed with distilled water, dried, and inserted into the test solution.
- After equilibration the pH is read and checked.
- Acid or alkaline error corrections are applied as necessary.
- The pH reading at the observed temperature is converted into that at a standard temperature,if needed. This can be done in several ways, including by use of a chart such as that in Fig. 7.2l.

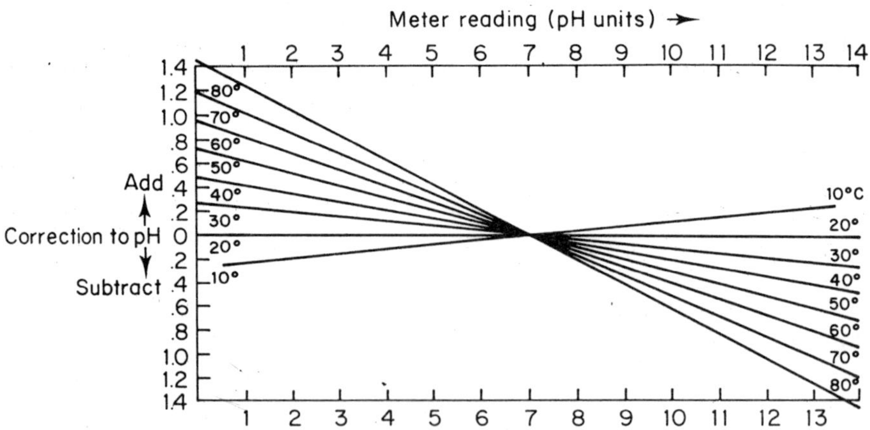

Fig. 7.2l. *pH temperature correction chart*

SAQ 7.2c

Before we leave pH meters and their probes let's revise some of the important practical points. Check your answers with mine and if you get them wrong or are unsure revise and recheck them.

(*a*) Name the four hydrogen-ion indicator electrodes and two main reference electrodes.

(*b*) Which hydrogen-ion indicator electrode is best to use:

(*i*) for strong alkali,
(*ii*) for convenience,
(*iii*) for an effluent from a dye works,
(*iv*) for a fairly uniform nearly neutral factory waste?

(*c*) What corrections would be applied when using a glass electrode to measure the pH of

(*i*) a sea water sample,
(*ii*) a blood sample,
(*iii*) a viscous polyphosphoric acid sample?

7.3. INDICATORS

A non-instrumental method of pH measurement much used in simple titrations is the use of indicators. These are organic dyes which change colour at or near the equivalence-point of a neutralisation to indicate the end of the reaction. As we shall see their usage is not quite as simple as that if good results are required.

7.3.1. Indicator Theory

As we have said already indicators are organic dyes which change colour as pH changes. This is because the indicator has two forms, one in acid and one in alkaline media. An example is Methyl Orange which in acid is (I), which is red. In alkali (II) is formed which is orange.

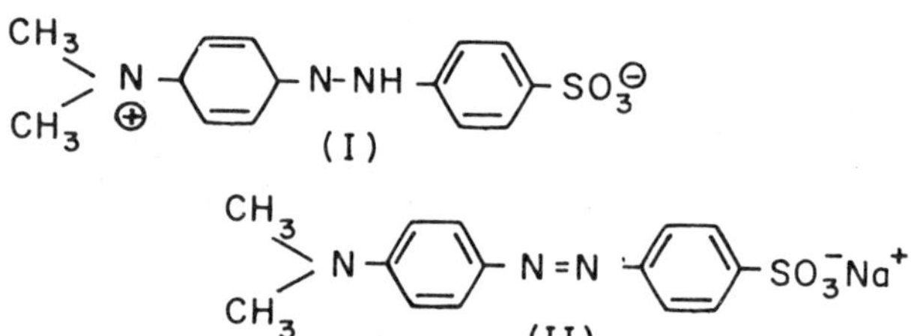

(I) (II)

If the colour change is clear and sharp enough then we can use the indicator as long as the pH change in the reaction cuts sharply across the pH range of the colour change of the indicator. This latter condition is not a simple matter to decide as will be seen later. In general we have two possible indicator reactions

(*a*) $HIn \rightleftharpoons H^+ + In^-$ ie it is a weak acid,

(*b*) $In + H_2O \rightleftharpoons InH^+ + OH^-$ ie it is a weak base

For these systems the Henderson–Hasselbach equation applies.

ie $pH = pK_a + \log \dfrac{[In^-]}{[HIn]}$ where pK_a is that of the indicator. The indicator will change colour at a pH near the pK_a value depending on the $[In^-]/[HIn]$ ratio.

Π Can you derive the above expression from the relevant equilibrium?

From $HIn \overset{K_a}{\rightleftharpoons} H^+ + In^-$, $K_a = \frac{[H^+][In^-]}{[HIn]}$

ie $\frac{1}{[H^+]} = \frac{[In^-]}{[HIn]} \times \frac{1}{K_a}$

∴ Taking logs

$$\log \frac{1}{[H^+]} = \log \frac{1}{K_a} + \log \frac{[In^-]}{[H^+]}$$

ie $pH = pKa + \log \frac{[In^-]}{[HIn]}$

In practice the pH range of an indicator is ±1 pH unit on either side of the pH corresponding to the pK_a value. This is because a change from when $[In^-]/[HIn]$ is 10 to when it is 1/10 is needed to register the colour change. This is the pH change necessary for the observer to notice a colour change and is not a function of how much indicator is used. The blank correction is the volume of acid or base required to neutralise the indicator itself.

Π What would be the indicator blank correction for say 0.2 cm^3 (6 drops) of 0.001*M*-methyl orange when used in a titration with 0.01*M*-NaOH?

The answer is 0.02 cm^3 of 0.01*M*-alkali as the indicator is 1/10th the concentration of the alkali.

Π Why does the presence of large amounts of a salt alter the pH range of an indicator?

Salts alter the indicator equilibrium in the direction away from the salt form towards the unionised form (if any). This will alter the $[In^-]/[HIn]$ ratio and hence the pH range.

7.3.2. Indicator Ranges

A list of the commonly encountered indicators with their pH ranges is given below in Fig. 7.3a, together with the colours of acid and base solutions. Notice that 'acid' and 'base' are relative, so that Bromophenol Blue, for example, is yellow below pH 3.0, and blue above pH 4.6 which is still acidic. Some indicators have a recognisable mid-shade. The Methyl Orange change is unclear, especially to those with partial colour blindness, but if a dye such as methylene

Indicator	Acid Colour	pH Range	Alkaline Colour
Cresol Red	red	0.2– 1.8	yellow
Thymol Blue	red	1.2– 2.8	yellow
Tropaeolin O	red	1.3– 3.0	yellow
Bromophenol Blue	yellow	3.0– 4.6	blue
Methyl Yellow	red	2.9– 4.0	yellow
Methyl Orange	red	3.1– 4.4	orange
Congo Red	blue	3.0– 5.0	red
Bromocresol Green	yellow	3.8– 5.4	blue
Methyl Red	red	4.2– 6.3	yellow
Bromocresol Purple	yellow	5.2– 6.8	purple
Bromophenol Red	yellow	5.2– 6.8	red
Bromothymol Blue	yellow	6.0– 7.6	blue
Phenol Red	yellow	6.8– 8.4	red
Thymol Blue	yellow	8.0– 9.6	blue
Phenolphthalein	colourless	8.3–10.0	red
Thymolphthalein	colourless	9.3–10.5	blue
Tropaeolin O	yellow	11.1–12.7	orange

Fig. 7.3a. *Colour changes and pH ranges of indicators*

blue or better still xylene cyanol FF is added to the Methyl Orange solution, then the colour change is from purple violet (acid) through grey to green. As we said before, the choice of indicator will depend partly on its pKa value, and its colour range, but most importantly on the type of neutralisation reaction.

7.4. NEUTRALISATION CURVES AND CHOICE OF INDICATORS

In this section we shall review neutralisation curves and the factors controlling the eventual choice of indicator for the reaction concerned.

7.4.1. Neutralisation Curves

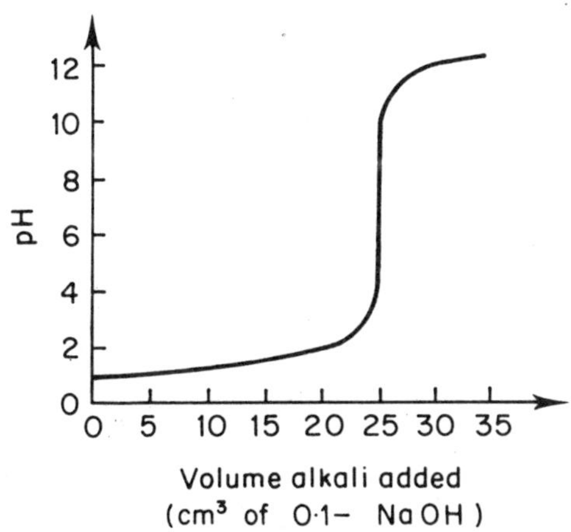

Fig. 7.4a. *Strong acid–strong alkali titration curve*

Fig. 7.4a shows a typical titration curve for a strong acid against a strong alkali. The significant feature is the very sharp and sudden change in pH near to the *equivalence-point* of the reaction. This is where the stoichiometric balance of the reaction is complete. This may or may not be at pH 7, depending on the degree of ionisation of the acid and base, ie their strengths as acid or base. Such a pH curve could be plotted either from the readings of a pH meter or by calculation. An example of such a calculation is given below.

Titration of 25.00 cm^3 of 0.1000 *M*-HCl *vs* 0.1000 *M* NaOH

Initially we have 25.00 cm^3 0.1000 *M* HCl with pH 1.00, then the following results:

Volume alkali (cm^3)	Total volume (cm^3)	[acid] or [base]	pH value
5.0	30.0	$20/30 \times 0.1 = 6.67 \times 10^{-2}$	1.18
10.0	35.0	$15/35 \times 0.1 = 4.29 \times 10^{-2}$	1.37
15.0	40.0	$10/40 \times 0.1 = 2.50 \times 10^{-2}$	1.60
20.0	45.0	$5/45 \times 0.1 = 1.11 \times 10^{-2}$	1.95
24.0	49.0	$1/49 \times 0.1 = 2.04 \times 10^{-3}$	2.69
24.5	49.5	$0.5/49.5 \times 0.1 = 1.01 \times 10^{-3}$	2.99
25.0	50.0	neutrality	7
25.5	50.5	$0.5/50.5 \times 0.1 = 9.90 \times 10^{-4}$ base	pOH = 3.00 pH = 11.00
30.0	55.0	$5/55 \times 0.1 = 9.09 \times 10^{-3}$ base	pOH = 2.04 pH = 11.96
40.0	65.0	$15/65 \times 0.1 = 2.31 \times 10^{-2}$ base	pOH = 1.64 pH = 12.36
50.0	75.0	$25/75 \times 0.1 = 3.33 \times 10^{-2}$ base	pOH = 1.48 pH = 12.52

These results when plotted on a graph give a curve as illustrated in Fig. 7.4a.

In practice with an unknown concentration of base or acid we would titrate roughly at first to find the approximate equivalence-point and then titrate more accurately.

If an indicator is used, we determine the end-point of the reaction by titration to the mid-tint of the indicator. This end-point will differ from the equivalence point for several reasons.

(*a*) the existence of the indicator blank ie the volume to titrate the indicator;

(*b*) the observer's choice of mid-tint;

(*c*) the pH range of the indicator, it needs a change of about 2 pH units for the colour change to occur.

The first item can be limited by using small volumes of indicator (a few drops) and by applying an indicator blank correction, and the second can be controlled but not eliminated by careful comparison with an agreed shade of colour. The third item cannot be eliminated and requires a pH shift of more than 2 units on the addition of one or two drops of acid or base. Hopefully if this addition takes us past the end-point and the equivalence-point (within this 2 pH unit change) at almost the same time, then the error is small and the two are effectively coincident.

If, however, there is no sharp pH change greater than or equal to 2 pH units or if the change does not coincide with the equivalent-point, then we may not be able to use indicators at all.

Π When else may an indicator not be useful?

Two possible answers are when the solution is strongly coloured itself or affects the dye used, and if the observer has sight problems such as colour blindness.

In our choice of indicators it is traditional to divide neutralisations into the following classes:

(*a*) strong acid *vs* strong base,
(*b*) strong acid *vs* weak base,
(*c*) weak acid *vs* strong base,
(*d*) weak acid *vs* weak base.

7.4.2. Graphical Measurement of Equivalence-point

Having obtained a pH curve, how do we process it to estimate the equivalence-point?

Reducing this to a purely mathematical problem, we have to determine the point of inflection where the slope changes. This is usually half-way on the near-vertical section of the pH change curve. It is assumed for the moment to be effectively coincident with the end-point and the equivalence-point.

It can be estimated

(*a*) by eye,
(*b*) by the tangent method below,
(*c*) by electronic or calculated use of change of pH with volume added, ie ΔpH per unit volume or by use of the second derivative.

We will consider method (*b*) now, and leave (*c*) till after potentiometric titrations.

A typical titration curve is shown below, as plotted from the results obtained by use of a pH meter. The initial and the final slope are drawn and produced as in Fig. 7.4b. Often the lines are nearly parallel. A perpendicular is drawn (hopefully to both lines) such that its mid-point is on the curve.

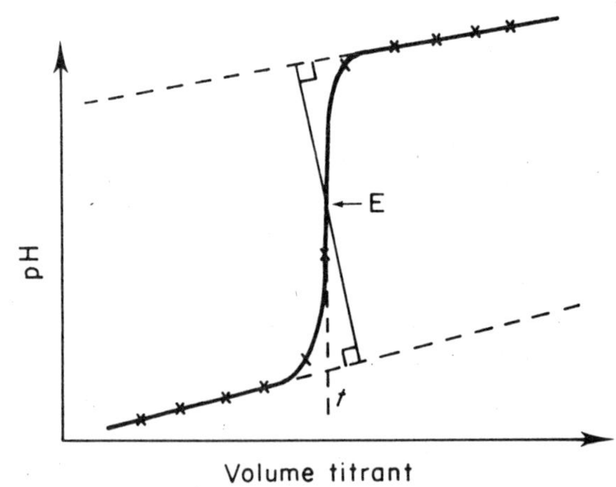

Fig. 7.4b. *Estimation by drawing of the equivalence point on a titration curve*

This is point E, giving a titration value t cm^3. Often the best results are obtained by careful plotting of several points within 5 cm^3 of the end point. Methods using the first or second derivative of the curve function will be dealt with in the section on potentiometric titration.

7.4.3. Strengths of Acids and Bases

The ionisation of acids and bases has been discussed elsewhere in Unit 2 in some detail, but I feel we should briefly review the situation to understand the choice of indicator and, later on, the theory and use of buffers.

The terms 'strong' and 'weak' are applied to the degree of ionisation in a given solvent (usually water) and are terms independent of 'concentrated' and 'dilute', although there is usually a mathematical relationship between them. In fact, with many partly ionised weak acids or bases, ionisation actually increases with dilution, so that the more dilute solution is more powerful or efficient as an ionic acid or base. Consequently the net $[H^+]$ or $[OH^-]$ is a function of acid or base strength and dilution or concentration. Using the Lowry–Brönsted concept, for acids we have, $HA + H_2O \rightleftharpoons A^- + H_3O^+$, and for bases, $B + H_2O \rightleftharpoons BH^+ + OH^-$.

$$K_a = \frac{[H_3O^+][A^-]}{[HA][H_2O]} \quad \text{and } K_b = \frac{[BH^+][OH^-]}{[B][H_2O]}$$

Water is acting as a proton acceptor (ie a base) in the first and a proton donor (ie an acid) in the second expression. Note that in either equation we have paired acids and bases (conjugate pairs) differing by a proton (or H_3O^+). So we have HA/A^-, H_2O/H_3O^+, B/BH^+, H_2O/OH^-, and more generally:

$$\text{acid 1} + \underset{(\text{base 2})}{\text{solvent}} \rightleftharpoons \text{base 1} + \text{acid 2}$$

$$\text{or base 1} + \underset{(\text{acid 2})}{\text{solvent}} \rightleftharpoons \text{acid 1} + \text{base 2}$$

From these expressions we can see how crucial is the role of the solvent in acting as one of the bases or acids.

Π How does water affect the following equilibria?

(*a*) $NaCl + H_2O \rightleftharpoons NaOH + HCl$
(*b*) $FeCl_3 + 3\,H_2O \rightleftharpoons Fe(OH)_3 + 3\,HCl$

This illustrates an alternative way of looking at neutralisation. Reaction (*a*) is usually termed the neutralisation of a strong base by a strong acid, whereas (*b*) is usually called salt hydrolysis. The difference lies in the action of the water which affects the position of equilibrium, (*a*) being extensively to the left and (*b*) more to the right. They are thus similar reactions with different equilibrium constants.

In our expressions for K_a and K_b, H_3O^+ is often replaced by H^+ and the water term is omitted as water is in great excess and effectively constant in concentration. Strong mineral acids have K_a values of $> 1 \times 10^{-1}$ and the values for weak acids range from 1×10^{-3} to 1×10^{-14} for the weakest acid in water, (ie water itself for which $K_a = K_w = 1 \times 10^{-14}$). Bases vary similarly in range.

Π If pK_a and pK_b are the negative logs of K_a and K_b as pH is of $[H^+]$, put the following pairs in order of strengths.

(*a*) acids, pK_a = (*i*) 3.23, (*ii*) 10.43,
(*b*) bases, pK_b = (*i*) 0.23, (*ii*) 4.36.

The answer in each case is that the former is the stronger as a K range of $10^{-1} \rightarrow 10^{-10}$ becomes a pK range of $1 \rightarrow 10$.

Degree of Ionisation

Strengths of acids and bases can be measured by the degree of ionisation, which can be calculated fairly easily. Take an acid of concentration, c, and concentration of ions, x.

$$HA + H_2O \rightleftharpoons H_3O^+ + A^-$$

	HA	H_3O^+	A^-
at the start	c	0	0
at equilibrium	$(c - x)$	x	x

ie $$K_a = \frac{x^2}{c - x}$$

and degree of ionisation $= \dfrac{x}{c} = \alpha$

This is a quadratic equation in x but if K_a is less than 1% of c, $K_a \simeq \dfrac{x^2}{c}$

So for ethanoic acid;

$$K_a = 1.76 \times 10^{-5} \text{ mol dm}^{-3}$$

If $c = 0.1M$ $\quad x = 1.33 \times 10^{-3}$ ie $\alpha = 1.3\%$

or for ammonia;

$$K_b = 1.75 \times 10^{-5} \text{ mol dm}^{-3}$$

If $c = 0.1M$ $\quad x = 1.33 \times 10^{-3}$ ie $\alpha = 1.3\%$.

Note the coincidence of values: both are about 1% ionised.

Π What is the percentage ionisation, if the concentration of the ethanoic acid is 1×10^{-3} M or 1×10^{-5} M?

Using $K_a = \dfrac{x^2}{c}$ and $K_a = 1.76 \times 10^{-5}$,

$$c = 0.001 \text{ or } 0.00001M$$

gives values of x of 1.33×10^{-4} and 1.33×10^{-5}

ie $\dfrac{x}{c} = 13.3\%$ and 133% respectively.

Note especially with the last answer the approximation condition that K_a is less than or equal to $0.01c$ is not true (hence odd answer).

$\therefore$ we should use $K_a = \dfrac{x^2}{c - x}$ ie $1.76 \times 10^{-5} = \dfrac{x^2}{1 \times 10^{-5} - x}$

ie $x^2 + (1.76 \times 10^{-5})x - 1.76 \times 10^{-10} = 0$

which gives $x = 0.71 \times 10^{-5}$ ie $\dfrac{x}{c} = 71\%$,

which confirms what we said about dilution and ionisation.

Similar calculation applies to bases,

ie $B + H_2O \rightleftharpoons BH^+ + OH^-$

$$K_b = \frac{[BH^+][OH^-]}{[B]}$$

$$K_b \approx \frac{x^2}{c} \text{ or precisely } = \frac{x^2}{(c - x)}$$

For a dibasic acid the maths calculation be slightly different:

$$H_2A \rightleftharpoons A^{2-} + 2H^+$$

$$\begin{array}{cccccl} c & & 0 & & 0 & \text{(initially)} \\ c - x & \rightleftharpoons & x & + & 2x & \text{(at equilibrium)} \end{array}$$

$$\therefore \quad K_a = \frac{2x^2}{c - x}$$

$$\approx \frac{2x^2}{c} \text{ if } K_a \leq 0.5\% \text{ of } c.$$

Usually we look at the ionisation of weak dibasic acids and bases in two steps, with K_{a1}, and K_{a2}.

7.4.4. Relationship Between K_a and K_w

We have already seen that salt hydrolysis is one way of looking at a neutralisation so that we can write

$$HA + H_2O \overset{K_a}{\rightleftharpoons} H_3O^+ + A^-$$

or

$$A^- + H_2O \overset{K_b}{\rightleftharpoons} HA + OH^-$$

where $K_a = \dfrac{[H_3O^+][A^-]}{[HA]}$ and $K_b = \dfrac{[HA][OH^-]}{A^-}$

Multiplying together, $K_a.K_b = [H_3O^+][OH^-]$

$= K_w$ for water, ie $K_a.K_b = K_w$

In a given solvent K_a for the acid and K_b for the conjugate base when multiplied together give the ionic product for the solvent. The practical consequence is that we can calculate K_a or K_b for a conjugate pair from a knowledge of the ionic product and K_b or K_a.

Π Calculate K_a for NH_4OH if $K_b = 1.75 \times 10^{-5}$ mol dm^{-3} at 25 °C.

As $K_a.K_b = 1 \times 10^{-14}$

$$K_a = \frac{1 \times 10^{-14}}{1.75 \times 10^{-5}} = 5.71 \times 10^{-10} \text{ mol dm}^{-3}$$

(or $pK_a = 9.24$)

Fig. 7.4c on the next page shows the pK_a values of some acids and bases.

Because of this salt hydrolysis and partial ionisation of acids and bases the pH of weak-acid or weak-base solutions will differ from that calculated from concentration alone; also salts of weak acids or bases will be hydrolysed giving non-neutral solutions. Let's take one example of each calculation.

A: ACIDS

Aliphatic	pK_a
methanoic (formic)	3.75
ethanoic (acetic)	4.76
propanoic	4.87
chloroethanoic (chloracetic)	2.86
lactic (2-hydroxypropanoic)	3.86
oxalic (ethanedioic)	1.27, 4.27
succinic (butane-1,4-dioic)	4.21, 5.64
maleic (*cis* but-2-ene-1,4-dioic)	1.92, 6.23
tartaric (2,3-dihydroxybutane 1,4-dioic)	3.03, 4.37
citric (3-carboxy-3-hydroxy pentane-1,5-dioic)	3.13, 4.76, 6.40
Aromatic	
benzoic	4.21
phthalic (Benzene-1,2-dicarboxylic)	2.95, 5.41
phenol	10.00
Inorganic	
boric	9.24
carbonic	6.37, 10.33,
phosphoric	2.12, 7.21, 12.32
sulphuric	−1, 1.92

B: BASES	pK_a
ammonia	9.24
methylamine	10.64
trimethylamine	9.80
ethylenediamine (1,2-diaminoethane)	7.00, 10.09
hydroxylamine	5.82
aniline	4.58
pyridine	5.17
hydrazine	7.93

Fig. 7.4c. *Dissociation constants of acids and bases at 25 °C in water (pK values)*

What is the pH of

(*a*) $0.100M\text{-}CH_3CO_2H$, $\qquad K_a = 1.76 \times 10^{-5}$,

(*b*) $0.100M\text{-}CH_3CO_2^-Na^+$, $\qquad K_a = 1.76 \times 10^{-5}$ for the acid?

(*a*) As before $\qquad CH_3CO_2H \rightleftharpoons CH_3CO_2^- + H^+$.
Let conc. be cM.

at equilibrium; $\qquad c - x \qquad x \qquad x$

$$\therefore \quad K_a = \frac{x^2}{(c - x)} \approx \frac{x^2}{c} \quad \therefore \quad x = 1.33 \times 10^{-3}M \text{ (as before)}$$

This is the $[H^+]$

$$\therefore \quad pH = \log \frac{1}{1.33 \times 10^{-3}} = 2.88$$

(If the acid were strong the pH would be 1.0)

(*b*) The reaction is:

$$CH_3CO_2^-\,Na^+ + H_2O \rightleftharpoons CH_3CO_2H + OH^-$$

at equilibrium;

$$(c - x) \qquad\qquad x \qquad x$$

$$\therefore \quad K_b = \frac{x^2}{(c - x)} \simeq \frac{x^2}{c}$$

$$K_b = \frac{K_w}{K_a} = \frac{1 \times 10^{-14}}{1.76 \times 10^{-5}} = 5.68 \times 10^{-10}$$

$$\therefore \quad x = (5.68 \times 10^{-10} \times 0.1)^{\frac{1}{2}}$$

$$= 7.54 \times 10^{-6}.$$ This is the $[OH^-]$,

ie a pOH of 5.12. $\therefore$ pH = 14 −pOH = 8.88,

ie not neutral.

Note that these are the pH values of single substances only, in water. For mixtures, say $CH_3CO_2H/CH_3CO_2Na^+$ (ie a buffer combination), the pH will differ, as both participate in the same equilibrium.

SAQ 7.4a

Calculate the pH at 25 °C of

(*a*) 0.100*M*-NH_4OH,

(*b*) 0.100*M*-NH_4Cl.

[$K_b(NH_4OH) = 1.76 \times 10^{-5}$ at 25 °C]

7.4.5. Titration Curves

In the light of this discussion of weak and strong acids and their salts we can return to titration curves

Strong Acid *vs* Strong Base

For this type of reaction (already discussed), the curve will have the shape we showed previously (Fig. 7.4a). We shall have pure acid at the start, gradually more dilute acid and increasingly concentrated neutral salt till at the equivalence-point we shall have neutral salt only. Immediately beyond we shall have salt plus increasing amounts of strong alkali. The change in pH near the equivalence-point will be sharp and large. There will be somewhat less of a change in pH if we use more dilute reagents as shown in Fig. 7.4d.

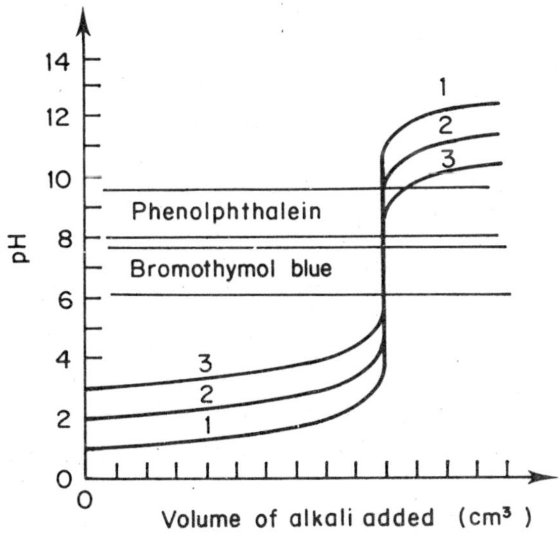

Fig. 7.4d. *Titration curves for strong acid (three different concentrations) titrations against strong alkali*

Curves 1, 2, and 3 are for successively smaller concentrations of strong acid. The change in pH at the equivalence point is about 7 pH units in 1, down to about 3.5 pH units in 3, with ranges of pH 10–3 and 5–8.5. When we look at our indicator ranges we find that Methyl Orange, Bromocresol Green, Methyl Red, Bromothymol Blue, Phenolphthalein, Litmus, and Cresol Red all fit comfortably for curve 1, whereas the choice is more restricted for curve 3, and Bromothymol Blue would do (as shown in Fig. 7.4d).

Π How would we cope with very dilute acid and base?

Both would have to be very dilute or we should get a very small titre, in fact we may have a pH range less than 2, so no indicator will do and potentiometric methods would have to be used.

Weak Acid *vs* Strong Base

If we titrated a weak acid such as ethanoic acid (acetic acid) against base we should get a curve such as that in Fig. 7.4e.

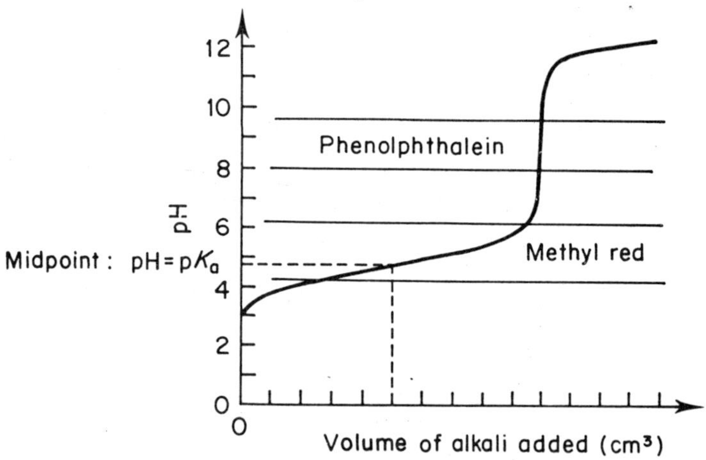

Fig. 7.4e. *Titration curve for a weak acid titrated against a strong base (NaOH)*

Here we start with a partly ionised acid with a solution of higher pH than for a strong acid, and as we neutralise it more acid ionises so we have mainly unionised acid + salt. Therefore the pH changes gradually. This is the buffer effect. At equivalence we shall have sodium ethanoate solution with a pH about 9 (rather than 7); we now get addition of alkali causing a sharp change in pH much as before. The vital pH change is about pH 6–pH 11, which means that indicators such as Phenolphthalein, Thymol Blue, and Cresol Red must be used.

Π What would happen if we used the wrong indicator such as Methyl Orange?

The colour change of the indicator would be diffuse, and occur over the addition of 2 or 3 cm^3 of titrant, which is clearly unsatisfactory.

Once again, dilution of titrants would restrict the pH range further.

The pH range of the reaction is also reduced as the acid strength of the weak acid is reduced as shown in Fig. 7.4f.

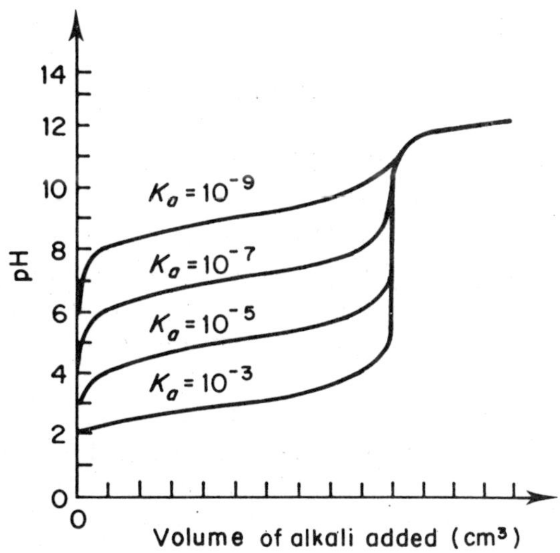

Fig. 7.4f. *Titration curves for the titration of a strong base(NaOH) against acids of differing strengths*

This means that very weak or very dilute acids may have to be titrated potentiometrically. Note also at half neutralisation, pH = pK_a (in Fig. 7.4e).

Strong Acid *vs* Weak Base

Predictably the converse of the above will hold for titration of a strong acid *vs* a weak base such as ammonia, so that the pH at equivalence is on the acid side and the curve is almost the mirror image of the previous one. The sharp change is from pH 3 to 7, so that suitable indicators are Methyl Red, Methyl Orange, or Bromocresol Green. The curve is shown in Fig. 7.4g.

As before, dilution or increasing weakness of base reduces the pH range still further as shown in Fig. 7.4h. A consequence is that indicator titration is usually done down to pK_a or pK_b of approximately 6 or 7, and potentiometric pH titration down to pK_a or pK_b of approximately 8 or 9.

Weak Acid *vs* Weak Base

Here the titration curve has a continuously and gradually changing pH with addition of base so that there is no region where a sharp shift in pH occurs for a small addition of titrant, or if there is, it is less than 2 pH units and so not detectable by using indicators. We have buffers on either side of the equivalence-point and at the equivalence-point the pH depends on the relative strengths of acid and base. As ammonia and ethanoic acid have identical K values in aqueous solution their salt would, in fact, be neutral.

We have two devices in this situation. Either plot a graph by using a pH meter, or alternatively change to another solvent which would change the reaction to strong acid *vs* weak base or strong base *vs* weak acid depending on the choice of solvent.

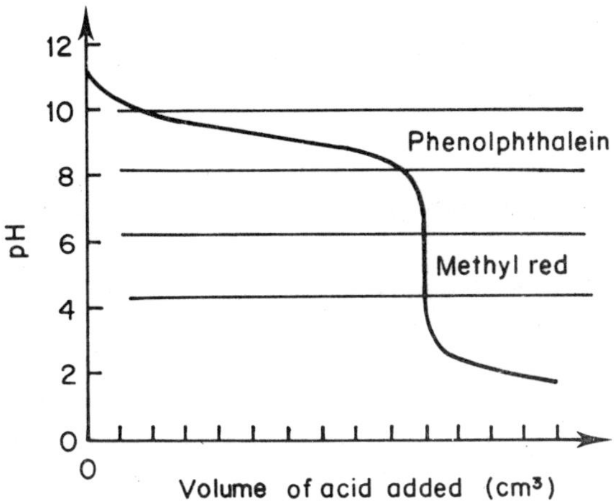

Fig. 7.4g. *Titration curve for the titration of a strong acid against a weak base*

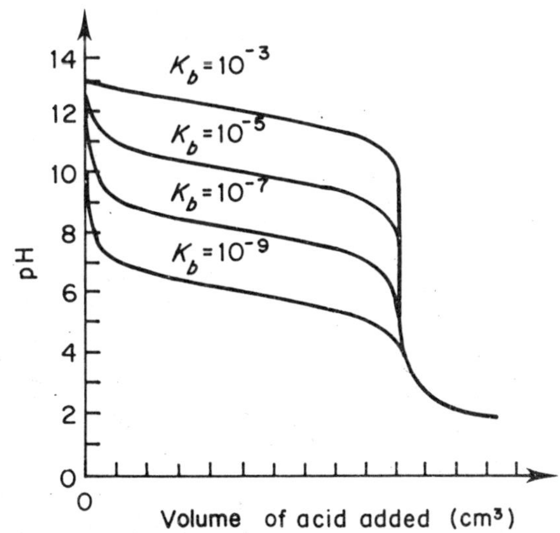

Fig. 7.4h. *Titration curves for the titration of a strong acid(HCl) against bases of differing strengths*

7.5. POLYBASIC ACIDS AND POLYACID BASES

Here we shall move on from monobasic acids and monoacid bases to polyprotic systems.

7.5.1. K_a and K_b for Polybasic Acids and Polyacid Bases

Fig. 7.5a below has a list of some of the common polyfunctional acids and bases with their K_a values. As more than one proton is involved, we have successive K_a values.

If we were to consider the individual ionisation steps of these acids and bases, we might consider H_2SO_4 and HSO_4^- as strong acids and $Ca(OH)_2$ and $Ca(OH)^+$ as strong bases, but all the rest count as weak acids or bases in water as solvent.

Remember too that here we have several linked equilibria, as for phosphoric acid, for example, we would have:

$$\underset{\text{(unionised)}}{H_3PO_4} \rightleftarrows H_2PO_4^- + H^+ \rightleftarrows HPO_4^{2-} + H^+ \rightleftarrows PO_4^{3-} + H^+$$

and theoretically we could have all these species present to one degree or another at any one time.

7.5.2. Neutralisation Curves for Polybasis Acids and Polyacid Bases

If we consider anew the titration curves for weak and strong acids against strong bases and add the two curves together sequentially we would get a two-stepped curve as in Fig. 7.5b.

The stronger acid is neutralised first along with some of the weaker acid (giving a curve more sloping than otherwise), and the sharp pH change of the strong acid still shows up although weaker here, and finally after all the strong acid has gone (equivalence-point A) we titrate the weak acid much as before (equivalence-point B).

Acid	Name	K_{a_1} (or K_{b_1})	pK_{a_1}	K_{a_2} or K_{b_2}	pK_{a_2}	K_{a_3} or K_{b_3}	pK_3
H_2SO_4	Sulphuric acid	>1	<-1	1.2×10^{-2}	1.92		
H_2CO_3	Carbonic acid	4.3×10^{-7}	6.37	4.7×10^{-11}	10.33		
H_3PO_4	Phosphoric acid	7.6×10^{-3}	2.12	6.2×10^{-8}	7.21	4.8×10^{-13}	12.30
HO_2CCO_2H	Ethan-1-2 dioic (oxalic acid)	5.4×10^{-2}	1.27	5.4×10^{-5}	4.27		
$H_2NCH_2CH_2NH_2$	1,2-Diaminoethane	1×10^{-7}	7.00	8.1×10^{-11}	10.09		
Na_2CO_3	Disodium carbonate	2.1×10^{-4}	3.7	2.3×10^{-8}	7.64		

(Data taken from Vogel's *Textbook of Quantitative Inorganic Analysis*, Longman, 1978)

Fig. 7.5a. *Polyfunctional acids and basis*

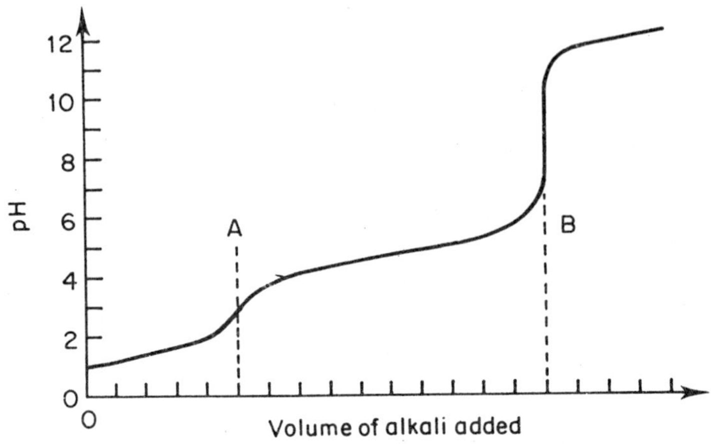

Fig. 7.5b. *Titration of a mixture of a strong and a weak acid with alkali*

Π Can we use indicators for A or B (in Fig. 7.5b)?

There isn't a sharp enough change at A (equivalence-point) so we can't pinpoint it with an indicator, but B is a typical weak acid strong base change, with pH changing sharply between pH 7 and pH 11 so that Phenolphthalein (and others) would be satisfactory.

Consider next the titration of carbonate with acid as shown in Fig. 7.5c.

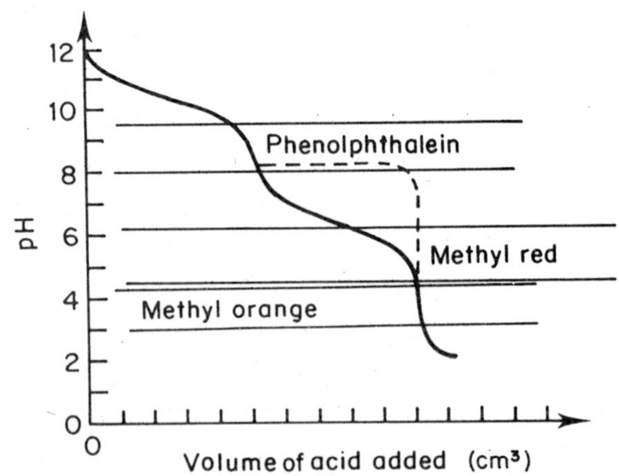

Fig. 7.5c. *Titration of carbonate with acid*

We have the following equilibria:

$$CO_3^{2-} + H^+ \rightleftharpoons HCO_3^-, \text{ and}$$

$$HCO_3^- + H^+ \rightleftharpoons H_2O + CO_2.$$

Initially we have the salt of a weak acid which reacts with, say, hydrochloric acid to give a mixture of carbonate and hydrogen carbonate ions, which buffer the pH change. Beyond the first equivalence-point the buffer is now hydrogen carbonate/free carbonic acid (dissolved CO_2), which buffers the solution past the second equivalence-point. Neither end-point has a sharp pH change and the use of Phenolphthalein (for the first) and Methyl Orange (for the second) is only approximate. If after the first end-point (with Phenolphthalein), we aerate or boil the solution to remove CO_2 gas, the graph follows the dotted line. Without the buffer effect a sharp pH change occurs at the second equivalence-point, where Methyl Orange or Methyl Red is satisfactory.

If you refer back to Fig. 7.5b and Fig. 7.5c you see that these are two-stage titrations, where the K values for each stage are quite different. If we took two strong acids such as hydrochloric acid and nitric acid they would be titrated together giving one (total) equivalence-point. The same is true for the two stages in the neutralisation of sulphuric acid which are titrated together (as both H_2SO_4 and HSO_4^- are strong acids). If we wish to obtain sodium hydrogen sulphate we have to titrate to the Phenolphthalein or Methyl Orange end point and then add a volume of acid equal to that first used.

In fact for a diprotic or triprotic compound we get clear separation of equivalence points only when the K_a (or K_b) values for each step show a ratio of $10^4:1$ (pK_a values differ by 4). Carbonate/bicarbonate only just does this (pK values 3.70 and 7.64) and ethanedioate (oxalate) does not meet this criterion (pK values: 1.27 and 4.27) and so the two stages cannot be distinguished. With phosphoric acid (pK_a values: 2.12, 7.21, 12.30) the three pK_a values are widely separated and should be clearly discernible. The third value is however, close to that of water (pK_w = 14) and is not normally discernible in this solvent. Because of the multi-buffer effect of polybasic ions few of the equivalence points are sharp, and pH-meter measurement is superior to use of indicators. However Phenolphthalein is often used to detect $H_3PO_4 \rightarrow H_2PO_4^-$, and Methyl Orange $H/_2PO_4^- \rightarrow HPO_4^{2-}$, although this is approximate.

7.5.3. Speciation in Polyacids

In our titration curves the equivalence-point marks the effective end of one species and the start of the increase of another and the pH/composition variation is often important. Later on we shall try to calculate the proportions of different ions at a given pH and the pH of mixtures of ions. This has importance outside titration curves as mixtures of polyprotic ions are commonly used in buffer solutions.

Π Are you clear, in broad terms, what a buffer is or does?

We shall be moving on to buffers in detail shortly, but as we have already mentioned buffering action in titration curves, we should be clear what we mean. A buffer is a mixture of salt + acid or base which can absorb the 'shock' of addition of strong acid or base and still keep the pH approximately the same. Hence the word 'buffer'.

Fig. 7.5d shows the calculated speciation of phosphate ions at different pH values. The vertical axis gives the proportion (or mole fraction) of each species and the most striking feature is that at any given pH, we have effectively only one or two out of four possible species present. For example, at pH 8 we have $H_2PO_4^-$ and HPO_4^{2-} (mainly), with no H_3PO_4 or PO_4^{3-}.

This arises, in fact, from the gap in pK_a values, so that corresponding to the stepwise neutralisation in the titration curve we get stepwise separation of species, which change as we move across the pH range. Fig. 7.5e shows a similar diagram for ethanedioate species.

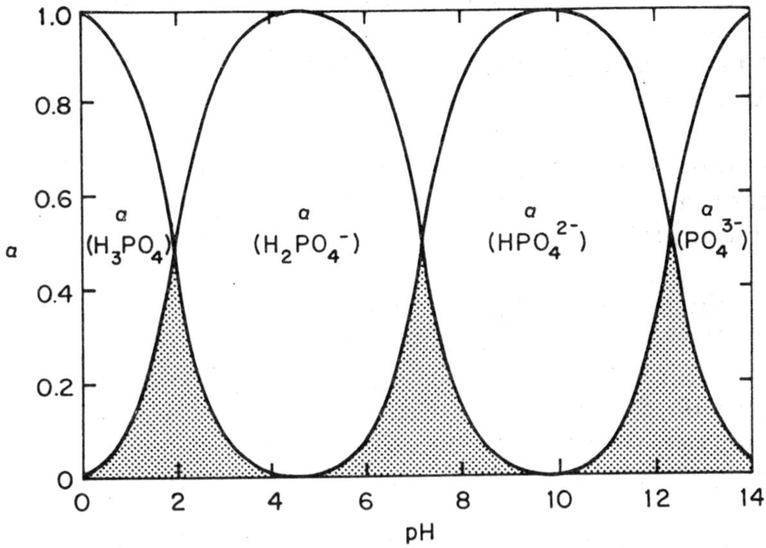

Fig. 7.5d. *Mole fraction of phosphate species as a function of pH*

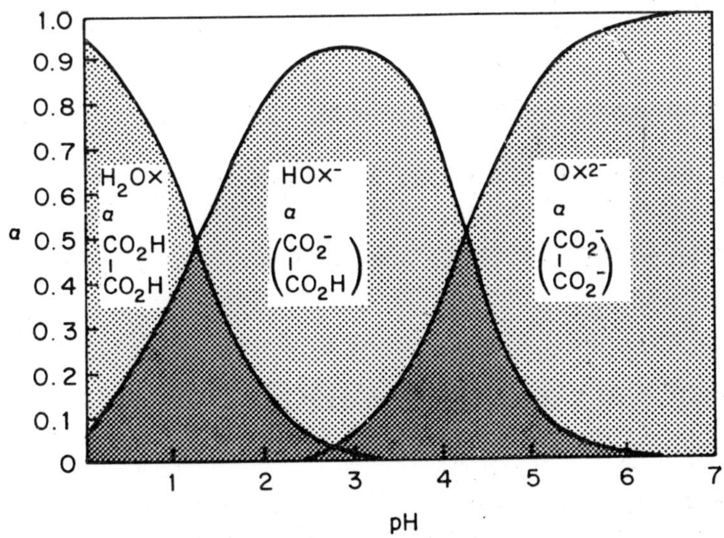

Fig. 7.5e. *Mole fraction of oxalate (ethanedioate) as a function of pH*

Here, as the titration curves overlap ($\Delta pK_a < 4$ units) we can have more than two species present, eg at pH 3 we have a small amount of H_2Ox [$(CO_2H)_2$], and mainly HOx^- [$CO_2H.CO_2^-$] and a small amount of Ox^{2-} [$^-O_2C.CO_2^-$].

As a consequence it is clear from the diagrams that only two salts (rather than three) are normally required to make a buffer solution of a given pH.

Π What would happen if we mixed three phosphate species say H_3PO_4, NaH_2PO_4, and Na_2HPO_4?

In fact with the mobility of the equilibria involved we should end up with two species. The outer two would be 'neutralised' to the other until one was used up.

Let us consider the phosphoric acid equilibria in more detail. We need to know two things:

(*a*) at a given pH which species are present?
(*b*) if we have one or more of these species in solution what is the pH of the solution?

7.5.4. Composition-pH Calculations

The possible equilibria we have for phosphate species are

$$H_3PO_4 + H_2O \rightleftharpoons H_3O^+ + H_2PO_4^-$$

$$K_{a1} = 7.6 \times 10^{-3} \qquad (pK_{a1} = 2.12)$$

$$H_2PO_4^- + H_2O \rightleftharpoons H_3O^+ + HPO_4^{2-}$$

$$K_{a2} = 6.2 \times 10^{-8} \qquad (pK_{a2} = 7.21)$$

$$HPO_4^{2-} + H_2O \rightleftharpoons H_3O^+ + PO_4^{3-}$$

$$K_{a3} = 4.8 \times 10^{-13} \qquad (pK_{a3} = 12.32)$$

or

$$PO_4^{3-} + H_2O \rightleftharpoons HPO_4^{2-} + OH^-$$

$$K_{b1} = 2.08 \times 10^{-2} \qquad (pK_{b1} = 1.68)$$

$$HPO_4^{2-} + H_2O \rightleftharpoons H_2PO_4^- + OH^-$$

$$K_{b2} = 1.61 \times 10^{-7} \qquad (pK_{b2} = 6.79)$$

$$H_2PO_4^- + H_2O \rightleftharpoons H_3PO_4 + OH^-$$

$$K_{b3} = 1.32 \times 10^{-12} \qquad (pK_{b3} = 11.88)$$

The first set are acid ionisation equilibria (protonation of water), and the second set, salt hydrolyses (protonation by water).

As we have shown before, $K_a.K_b = K_w$ for a conjugate pair,

ie $\quad K_{a1}.K_{b3} = K_w$

$= K_{a2}.K_{b2} = K_{a3}K_{b1}$

Let us try to calculate:

(*a*) the pH of solutions of individual phosphate species,
(*b*) the pH of solutions of mixtures of two phosphate species.

Example 1. What is the pH of 0.1000 M-H_3PO_4?

$$H_3PO_4 + H_2O \overset{K_{a1}}{\rightleftharpoons} H_3O^+ + H_2PO_4^-$$

initially	0.100	0	0
at equilibrium	$0.100 - x$	x	x

$$\therefore \quad K_{a1} = \frac{x^2}{(0.100 - x)} = 7.6 \times 10^{-3}$$

Here we *cannot* use the approximation as the concentration is not $100 \times K_a$.

$$\therefore \quad x^2 + 7.6 \times 10^{-3}x - 7.6 \times 10^{-4} = 0$$

$$x = -\frac{7.6 \times 10^{-3} \pm [(7.6 \times 10^{-3})^2 + 4 \times 1 \times 7.6 \times 10^{-4}]^{\frac{1}{2}}}{2}$$

$$= -\frac{7.6 \times 10^{-3} \pm 5.57 \times 10^{-2}}{2} = 2.41 \times 10^{-2}$$

(taking + value)

This gives a pH of 1.62.

Example 2. What is the pH of 0.1000 M-Na_3PO_4?

By using the salt hydrolysis equation we get

$$PO_4^{3-} + H_2O \underset{}{\overset{K_{b1}}{\rightleftharpoons}} HPO_4^{2-} + OH^-$$

	PO_4^{3-}	HPO_4^{2-}	OH^-
initially	0.100	0	0
at equilibrium	$0.100 - x$	x	x

$$\therefore \quad K_{b1} = \frac{x^2}{(0.100 - x)} = 2.08 \times 10^{-2}$$

Once again we cannot use the approximation ($c < 100\ K_b$).

$$\therefore \quad x^2 + 0.0208x - 0.00208 = 0$$

Solving this gives $x = 0.036M = [OH^-]$

$$\therefore \quad [H^+] = \frac{1.0 \times 10^{-14}}{0.036} = 2.8 \times 10^{-13}$$

Hence pH = 12.55

SAQ 7.5a

What is the pH at 25 °C of 0.0200 M-$Na_2C_2O_4$?

($K_{a2} = 5.4 \times 10^{-5}$
$K_w = 1.00 \times 10^{-14}$)

Now let's consider an amphoteric salt such as NaH_2PO_4. It is amphoteric in that it can act both as a proton donor or acid (giving HPO_4^{2-}) or proton-receiving base giving H_3PO_4.

Example 3. What is the pH of 0.100 *M*-$NaHCO_3$?

Amphoteric acids or bases produce complicated arithmetic. We have a competition between ionisation producing H^+;

ie $HA^- + H_2O \overset{K_a}{\rightleftharpoons} H_3O^+ + A^{2-}$

and base hydrolysis producing OH^-

ie $HA^- + H_2O \overset{K_b}{\rightleftharpoons} H_2A + OH^-$

The two *K* values control the amounts of OH^- and H^+ produced and the resultant pH. It can be shown for such a system that;

$$[H^+] = \left[\frac{K_{a2}[HA^-] + K_w}{x^1 + ([HA^-]/K_{a1})}\right]^{\frac{1}{2}}$$

which simplifies to $[H^+] = [K_{a1}.K_{a2}]^{\frac{1}{2}}$

if

$$K_2\,[HA^-] \gg K_w$$

and

$$[HA^-]/K_{a1} \gg 1.$$

For 0.100 *M*-$NaHCO_3$;

$K_{a1} = 4.3 \times 10^{-7}$
$K_{a2} = 4.7 \times 10^{-11}$

$\therefore \quad K_{a2}\,[HA^-]$ is $\gg K_w$

and $[HA^-]/K_{a1} \gg 1$

$$\therefore \quad [H^+] = [K_{a1}K_{a2}]^{\frac{1}{2}} = [4.3 \times 10^{-7} \times 4.7 \times 10^{-11}]^{\frac{1}{2}}$$

$$= 4.50 \times 10^{-9}, \quad \text{ie pH} = 8.35$$

SAQ 7.5b **What is the pH at 25 °C of 0.100 M-Na_2HPO_4?**

Now let's move on to mixtures.

Example 4.

What is the pH of 0.2000 M-NaH_2PO_4/0.1000 M-Na_2HPO_4?

To simplify the calculation let us assume the contributions from Na_3PO_4 and H_3PO_4 are negligible (as shown by distributions in Fig. 7.5d due to the large differences in magnitude of K_{a1} K_{a2} and K_{a3}).

$$H_2PO_4^- + H_2O \overset{K_{a2}}{\rightleftharpoons} H_3O^+ + HPO_4^{2-} \quad K_{a2} = 6.2 \times 10^{-8}$$

	$H_2PO_4^-$	H_3O^+	HPO_4^{2-}
initially	$0.2000M$	0	$0.1000M$
at equilibrium	$0.2 - x$	x	$0.1 + x$

$$\therefore \quad K_{a2} = \frac{(x) \times (0.1 + x)}{(0.2 - x)} \simeq \frac{0.1x}{0.2}$$

as either concentration $> 100\ K_{a2}$

$\therefore \quad x = 1.24 \times 10^{-7}$ Hence pH $= 6.91$.

SAQ 7.5c

> **What is the pH at 25 °C of 0.1000 M-H_3PO_4/0.0200 M-NaH_2PO_4?**
>
> **($K_{a1} = 7.6 \times 10^{-3}$)**

Now let us turn our attention to the reverse problem. Given the pH of the mixture can we calculate its composition? We need to set up a series of equations connecting pH and the concentration of each of the phosphate (or ethanedioate) species. With three unknowns we need four equations. The fourth equation comes from the total mass balance in that total [phosphate] = sum of all [phosphate species]

Suppose we represent the fraction of each species as a proportion of the total phosphate so that;

$$\alpha_0 = \frac{[H_3PO_4]}{[\text{total } PO_4]} \qquad \alpha_1 = \frac{[H_2PO_4^-]}{[\text{total } PO_4]}$$

$$\alpha_2 = \frac{[HPO_4^{2-}]}{[\text{total } PO_4]} \qquad \alpha_3 = \frac{[PO_4^{3-}]}{[\text{total } PO_4]}$$

Hence

$$\alpha_0 + \alpha_1 + \alpha_2 + \alpha_3 = 1 \qquad (i)$$

(*i*) α's are the mole fractions of each species in terms of total phosphate.

Using the acid dissociation equilibria we get three equations.

$$[PO_4^{3-}] = \frac{K_{a3}[HPO_4^{2-}]}{[H^+]} \qquad (ii)$$

$$[HPO_4^{2-}] = \frac{K_{a2}[H_2PO_4^-]}{[H^+]} \qquad (iii)$$

$$[H_2PO_4^-] = \frac{K_{a1}[H_3PO_4]}{[H+]} \qquad (iv)$$

Lets eliminate all terms except $[H^+]$, [total PO_4], and say

$[H_2PO_4^-]$ Eq. (*iii*) remains, ie

$$[HPO_4^{2-}] = K_{a2}\frac{[HPO_4^-]}{[H^+]}$$

Substituting from (*iii*) into (*ii*) gives;

$$[PO_4^{3-}] = \frac{K_{a3}K_{a2}[H_2PO_4^-]}{[H^+]^2}$$

(*iv*) becomes $[H_3PO_4] = \dfrac{[H^+][H_2PO_4^-]}{K_{a1}}$

and substituting into Eq. (*i*) for each species gives

$$[\text{total } PO_4] = \frac{[H^+][H_2PO_4^-]}{K_{a1}} + [H_2PO_4^-]$$

$$+ \frac{K_{a2}[H_2PO_4^-]}{[H^+]} + \frac{K_{a2}K_{a3}[H_2PO_4^-]}{[H^+]^2}$$

Now divide by $[H_2PO_4^-]$. This gives:

$$\frac{[\text{total } PO_4]}{[H_2PO_4^-]} = \frac{[H^+]}{K_{a1}} + 1 + \frac{K_{a2}}{[H^+]} + \frac{K_{a2}K_{a3}}{[H+]}$$

$$= \frac{[H^+]^3 + K_{a1}[H^+]^2 + K_{a1}K_{a2}[H^+] + K_{a1}K_{a2}K_a3}{K_{a1}[H^+]^2}$$

But $$\frac{[\text{total } PO_4]}{[H_2PO_4^-]} = \frac{1}{\alpha_1}$$

Hence

$$\alpha_1 = \frac{K_{a1}[H^+]^2}{([H^+]^3 + K_{a1}[H^+]^2 + K_{a2}K_{a1}[H^+] + K_{a1}K_{a2}K_{a3})}$$

Similarly

$$\alpha_0 = \frac{[H^+]^3}{([H^+]^3 + K_{a1}[H^+]^2 + K_{a2}K_{a1}[H^+] + K_{a1}K_{a2}K_{a3})}$$

$$= \frac{[H^+]^3}{D}$$

$$\alpha_2 = K_{a1}K_{a2}[H^+]/D$$

$$\alpha_3 = K_{a1}K_{a2}K_{a3}/D$$

where $D = [H^+]^3 + K_{a1}[H^+]^2 + K_{a1}K_{a2}[H^+] + K_{a1}K_{a2}K_{a3}$

(N.B $\alpha_0 + \alpha_1 + \alpha_2 + \alpha_3 =$

$[H^+]^3/D + K_{a1}[H^+]^2/D + K_{a1}K_{a2}[H^+]/D + K_{a1}K_{a2}K_{a3}/D = 1$)

This looks very tedious (and it is somewhat). Let's work out one example noting first that the denominator, $D = [H^+]^3 + K_{a1}[H^+]^2$ etc, is the same for all and that terms are repeated conveniently.

Example 5.

0.100 M-H_3PO_4 is neutralised to pH 6.0; what species are present and in what percentage?

$[H^+] = 1.00 \times 10^{-6}$ $K_{a1} = 7.6 \times 10^{-3}$
$K_{a2} = 6.2 \times 10^{-8}$
$K_{a3} = 4.8 \times 10^{-13}$

Hence

$[H^+]^3 = 1.00 \times 10^{-18}$ $K_{a1}[H^+]^2$
$= 7.6 \times 10^{-3} \times 10^{-12} = 7.6 \times 10^{-15}$

$K_{a1}K_{a2}[H^+] = 7.6 \times 10^{-3} \times 6.2 \times 10^{-8} \times 10^{-6}$
$= 4.71 \times 10^{-16}$

$K_{a1}K_{a2}K_{a3} = 7.6 \times 10^{-3} \times 6.2 \times 10^{8} \times 4.8 \times 10^{-13}$
$= 2.26 \times 10^{-22}$

and

$[H^+]^3 + K_{a1}[H^+]^2 + K_{a1}K_{a2}[H^+] + K_{a1}K_{a2}K_{a3} = 8.07 \times 10^{-15}$

Hence

$$\alpha_0 = \frac{1 \times 10^{-18}}{8.07 \times 10^{-15}} = 1.24 \times 10^{-4}$$

$$\alpha_1 = \frac{7.6 \times 10^{-15}}{8.07 \times 10^{-15}} = 9.4 \times 10^{-1}$$

$$\alpha_2 = \frac{4.72 \times 10^{-16}}{8.07 \times 10^{-15}} = 5.85 \times 10^{-2}$$

$$\alpha_3 = \frac{2.26 \times 10^{-22}}{8.07 \times 10^{-15}} = 2.80 \times 10^{-8}$$

Hence concentrations are

$c \times \alpha_0 = 1.24 \times 10^{-5}$ M-H_3PO_4, (about 10^{-2}%)

$c \times \alpha_1 = 9.4 \times 10^{-2}$ M-$H_2PO_4^-$ (94%)

$c \times \alpha_2 = 5.85 \times 10^{-3}$ M-HPO_4^{2-} (5.85%)

$c \times \alpha_3 = 2.80 \times 10^{-9}$ M-PO_4^{3-} (about 10^{-6}%)

Hence percentages are	H_3PO_4	$H_2PO_4^-$	HPO_4^{2-}	PO_4^{3-}
	trace	94	6	trace

This fits in nicely with the pH distribution curves of Fig. 7.5c.

SAQ 7.5d

> What if the pH of the neutralised 0.100 M-H_3PO_4 was now 8.0?
>
> ($H^+ = 1.0 \times 10^{-8}$ $K_{a1} = 7.6 \times 10^{-3}$
>
> $K_{a2} = 6.2 \times 10^{-8}$ $K_{a3} = 4.8 \times 10^{-13}$).

SAQ 7.5d

7.6. BUFFER SYSTEMS

We have, during our discussion of titration curves and elsewhere had some occasion to mention buffers and buffer action. Here we shall go into it in a little more detail.

7.6.1. Theory of buffers

Before we explore buffer action let us look again at weak acids.

$$HA \overset{K_a}{\rightleftharpoons} H^+ + A^-$$

Hence $K_a = \dfrac{[H^+][A^-]}{[HA]}$ and $\dfrac{1}{[H^+]} = \dfrac{1}{K_a}\dfrac{[A^-]}{[HA]}$

Taking logs $\log \dfrac{1}{[H^+]} = \log \dfrac{1}{K_a} + \log \dfrac{[A^-]}{[HA]}$

ie
$$pH = pK_a + \log \frac{[A^-]}{[HA]}$$

This is often called the buffer equation, more formally it is know as the *Henderson–Hasselbach* equation. Let's examine its consequences. It is an equation similar to the indicator equation and as before, we now consider the effect of changing the $[A^-]/[HA]$ ratio.

If the ratio of $[A^-]$ to $[HA]$ is 10 : 1 $pH = pK_a + 1$ and if it is 1 : 10 $pH = pK_a - 1$. Hence for a hundred-fold change in ratio only a small change (2 pH units) in pH occurs. Thus if we arrange to have relatively large and approximately equal concentrations of A^- and HA we shall have a buffer system whose pH can change significantly only with very large shifts in the A^-/HA system

Π Can you derive the same equation for weak bases?

$$K_b = \frac{[BH^+][OH^-]}{[B]}$$

$$\therefore \quad \log \frac{1}{[OH^-]} = \log \frac{1}{K_b} + \log \frac{[BH^+]}{[B]}$$

$$\text{ie} \quad pOH = (14 - pH) = pK_b + \log \frac{[BH^+]}{[B]}$$

ie only small changes in pOH (or pH) for large $[BH^+]/[B]$ changes.

Notice also the salt effect or common-ion effect here.

The pH of the solution depends strongly on the salt of the acid or base, and salt extra to that provided by the acid or base will repress ionisation and alter the pH of the solution.

SAQ 7.6a

Calculate the pH of

(*a*) 0.0100 M-CH_3CO_2H,

($K_a = 1.76 \times 10^{-5}$)

(*b*) 0.0100M-CH_3CO_2H/0.100M.$CH_3CO_2^-Na^+$

SAQ 7.6a

Thus to make a buffer, the major recipe is to take a weak acid and its salt or take a weak base and its salt. We also observe that the pH of such mixtures will generally lie in the region of pK_a ± 2 units.

Samples of buffers include the following

Acid/Salt	Typical pH value (pK_a)
CH_3CO_2H/CH_3CO_2Na	4.75
Phthalic Acid/KHP	2.89
KHP/Potassium phthalate	5.41
Na_2HPO_4/NaH_2PO_4	7.21
Na_2HPO_4/Na_3PO_4	12.32
$NaHCO_3/Na_2CO_3$	10.25

Base/Salt	Typical pH value (pK_a)
NH_4OH/NH_4Cl	9.26

[KHP = potassium hydrogen phthalate]

A sketch may illustrate buffer action.

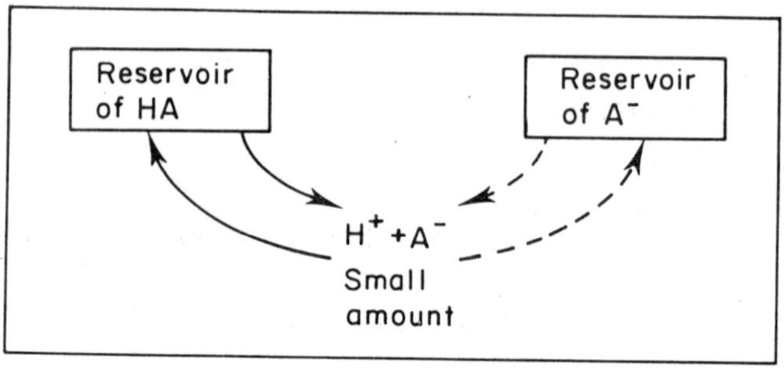

(*a*) Addition of HX.

The new H^+ will react with the A^- to produce HA (increase in reservoir of HA, decrease in that of A^-), $[H^+]$ much as before.

(*b*) Addition of NaOH.

H^+ neutralised, fresh H^+ generated from HA (decrease in reservoir of HA, increase in that of A^-)

A similar sketch can be drawn for a base/salt system.

Another buffer type is the *amphoteric salt* as discussed in the last section. Salts such as Na_2HPO_4, NaH_2PO_4, KH phthalate, and $NaHCO_3$ have a set of pH values almost independent of concentration, but have far less capacity for compensating for addition of acid or base. They have a low buffer capacity. They are, however, useful as a standard of pH for pH meters.

7.6.2. Buffer Calculations

The pH of a buffer solution can always be worked out from its composition, as can the composition from the pH value in simple cases but, as the calculations are often tedious, when a buffer of a given pH is needed one is selected from a compilation such as that in Fig. 7.6a.

pH at 25 °C	cm^3 0.2 *M* KCl	cm^3 0.1 *M* KHPhthalate	cm^3 0.1 *M* KH_2PO_4	cm^3 0.2 *M* HCl	cm^3 0.1 *M* HCl	cm^3 0.1 *M* NaOH	cm^3 0.025 substrate	cm^3 0.05 *M* Na_7HPO_4	cm^3 0.2 *M* NaOH
1.0	25.0			67.0					
2.0	25.0			6.5					
3.0		50.0			22.3				
4.0		50.0			0.1				
5.0		50.0				22.6			
6.0			50.0			5.6			
7.0			50.0			29.1			
8.0			50.0			46.1			
9.0					4.6		50.0		
10.0						18.3	50.0		
11.0						4.1		50.0	
12.0						26.9		50.0	
13.0	25.0								66.0

Fig. 7.6a. *Recipe for buffers of various pHs*

Buffer calculations have been incidentally performed in the section on speciation in polyacids and composition-pH calculations, and so I give you just one more example here on which to practise.

SAQ 7.6b

(*a*) What is the pH of 1.00 *M*-Na_2CO_3/1.00 *M*-$NaHCO_3$ solution?

(*b*) What would be the mass of each in 1.00 dm^3 to give a buffer of pH 9.0?

Data $K_{b1} = 2.1 \times 10^{-4}$ $pK_{b1} = 3.7$

7.6.3. Buffer Capacity

As we have just seen the *ratio* of salt concentration to weak-acid or weak-base concentration determines the pH. What we haven't considered in detail is the capacity of the buffer to absorb the effects of acid or base.

In the bicarbonate–carbonate buffer above, carbonate counters acid addition and bicarbonate base addition, and as in simple titration this neutralising power is a function of the molarity of the component. So the buffer solution of pH 10.3, if 0.1 *M* in both substances is at the mid-point of its action and is better than either a 0.01 *M* solution of each or a 0.1 *M* solution of one and an 0.01 *M* solution of the other. Good practice will require us to use a buffer and at its optimum pH (with equal concentration of A^- and HA and with sufficient capacity (ie joint molarity) for our needs.

7.6.4. Use of Buffers

As we have said buffers are most effective near the pH corresponding to the pK_a or pK_b value of one of the components, although it is possible to make buffer solutions of almost any pH by use of suitable mixtures. In practice, some of these buffers have little capacity and are used essentially for pH calibration. Note that amphoteric salts at most concentrations have a constant pH but little buffering power.

Buffer solutions can also be made up to a recipe for constant ionic strength, for total molarity, or for the molarity of any ion. Constant ionic strength is important in biological systems, where osmotic and other effects would occur if this was not so.

Amongst the common uses of buffer solutions are the following.

- Biological buffers: HCO_3^-/CO_3^{2-} giving a pH of 7.4 at 37 °C is the major buffer in blood. Phosphate buffers are common although their action may be aided by that of amino-acids. Enzyme activity and the related reaction rate are often pH-dependent and buffering is essential. Antacid tablets tend to buffer stomach acids.

- Calibration of pH solutions, eg with a pH meter.

- With ion-selective electrodes, where $[H^+]$ or $[OH^-]$ needs to be diminished or kept constant to prevent interference (eg TISAB buffer for F^- electrodes)

- For electroplating and similar solutions where pH control is essential.

Buffers do have disadvantages as follows.

- They have a finite buffering power and limited pH range.
- The ionic content of the solutions is high.
- The ions of a buffer may precipitate, inhibit, complex with, or otherwise react with the analytes of interest.
- Some buffers are growth-promoting for micro-organisms, and algae and bacteria may therefore cause problems.

7.7. NON-AQUEOUS SOLVENTS

On several occasions we've mentioned non-aqueous solvents, so I thought it would be useful to gather some information about their use in neutralisation reactions.

7.7.1. Classification of Solvents

Fig. 7.7a gives a list of the potentially usable non-aqueous solvents.

Amphiprotic solvents are those capable of acting both as a Lowry–Brönsted acid and as a Lowry–Brönsted base (and both simultaneously).

Hence

$$\begin{array}{rcl} & & \text{acid} \quad\quad \text{base} \\ 2\,H_2O & \rightleftharpoons & H_3O^+ + OH^- \\ 2\,NH_3 & \rightleftharpoons & NH_4^+ + NH_2^- \\ 2\,CH_3CO_2H & \rightleftharpoons & CH_3CO_2H_2^+ + CH_3CO_2^- \\ 2\,C_2H_5OH & \rightleftharpoons & C_2H_5OH_2^+ + C_2H_5O^- \end{array}$$

Solvent	Classification
Acetic acid (Ethanoic acid) (CH_3COOH)	Amphiprotic – acidic
Acetone (Propanone) (CH_3COCH_3)	Aprotic – neutral
Acetonitrile (Methyl cyanide) (CH_3CN)	Aprotic – neutral
Ammonia (NH_3)	Amphiprotic – basic
Chloroform (Trichloromethane) ($CHCl_3$)	Aprotic – neutral
Dimethylformamide [$HCON(CH_3)_2$]	Aprotic – basic
Dimethyl sulphoxide (CH_3SOCH_3)	Aprotic – basic
Ethanol (C_2H_5OH)	Amphiprotic – neutral
Ethylenediamine (1,2-Diaminoethane) ($H_2NCH_2CH_2NH_2$)	Amphiprotic – basic
Formic Acid (Methanoic acid) ($HCOOH$)	Amphiprotic – acidic
Methanol (CH_3OH)	Amphiprotic – neutral
Methyl Ethyl Ketone (Butanone) ($CH_3COC_2H_5$)	Aprotic – neutral
Methyl Isobutyl Ketone (4-Methylpentan-2-one) [$(CH_3)_2CHCH_2COCH_3$]	Aprotic – neutral
Phenol (C_6H_5OH)	Amphiprotic – acidic
Pyridine (C_5H_5N)	Aprotic – basic
Water (H_2O)	Amphiprotic – neutral

Fig. 7.7a. *Classification of solvents for non-aqueous titration*

These can be ranked (by their ease of accepting a proton from water, for example), and hence we can classify them as amphiprotic, acidic, etc. Hence CH_3CO_2H is amphiprotic-acidic, but NH_3 is amphiprotic-basic. Note, however, that water alone cannot be the standard as NH_2^- and $C_2H_5O^-$ cannot exist in aqueous solution.

Π Why not?

They are too basic and would abstract a proton from water immediately to give NH_3 or C_2H_5OH. For the same reason OH^- is unlikely to exist in large amounts in CH_3CO_2H solution. Notice also the usual pK_b for ammonia refers to the equilibrium below,

$$NH_3 + H_2O \rightleftharpoons NH_4^+ + OH^-$$

$$(\textit{not} \text{ to } NH_2^- + H_2O \rightleftharpoons NH_3 + OH^-).$$

Aprotic solvents do not release protons, but may act either as a simple solvent, where polarity as measured by the dielectric constant is significant, or they may act as a proton acceptor ie aprotic basic. There are no known aprotic-acidic solvents. Hence pyridine is aprotic-basic (giving pyridine-H^+). Many acids and bases of interest are intrinsically more soluble in these solvents, having low solubility in water. A more fundamental reason for the use of non-aqueous solvents lies in the nature of the titration curve. A strong acid is only such when it protonates all of the equivalent number of solvent molecules in that solvent. If we use a solvent which is harder to protonate, then our particular acid may only partially ionise, ie be a weak acid. Conversely a weak acid may ionise in a more basic solvent enough to become a strong acid. For example, HCl in water is a strong acid but in ethanoic acid is a weak acid. Notice that all acids capable of 100% ionisation in a given solvent are equally strong in that solvent, and cannot be distinguished by strength. This is the so-called levelling effect.

An idea of the shift in acid-strength when the solvent is changed can be gained from the data below (Fig. 7.7b).

	Ethanoic acid	Water	Liquid ammonia
Strong acid	$HClO_4$	$HClO_4$ HNO_3 H_2SO_4 HCl	$HClO_4$ HNO_3 HCl CH_3CO_2H
Weak Acid	HCl HNO_3 H_2SO_4	CH_3CO_2H	H_2O
'Neutrality'	CH_3CO_2H	H_2O	NH_3
Weak base	H_2O	NH_4OH	NaOH
Strong base	NH_4OH	NaOH, $NaNH_2$	$NaNH_2$

Fig. 7.7b. *Strengths of acids and bases in different solvents*

7.7.2. Non-aqueous Titrations

Non-aqueous titrations have most of the features of aqueous titrations but with some modifications. Anhydrous conditions are often necessary (unless titrating water itself), as water is a complicating acid or base in the non-aqueous solvent. pH, K_a and K_b do not have the same values as for aqueous solutions, as all are based on equilibria in water, but parallel concepts exist. Indicators used in one solvent cannot necessarily be transferred to another. The reaction of pH probes will also be different, as different equilibria exist externally to the probe, but pH meters do work well with most non-aqueous solvents.

In aprotic solvents, the usual solvent-base and solvent-acid reactions may not occur; alternatively we may have ion-pairs formed. For example, in ketone solvents or alcohols, pyridine and an acid may be present as an ion pair, ie Pyridine-H^+-base$^-$, rather than as a salt plus water.

Experimentally, we use a pH meter with a reference electrode (often modified) and record the changing potential as titrant is added. Indicators such as Methyl Violet (violet → blue → green) can be used. The equivalence-point can be found from the titration curve as described before and in the next section. Standardisation may require different primary standards; one of wide application is potassium hydrogen phthalate.

A typical non-aqueous titration might be that of aniline *vs* perchloric acid in *anhydrous* ethanoic acid (usually ethanoic anhydride is added to ensure this) with Methyl Violet as indicator. We might also titrate benzoic acid in methanol–benzene by using sodium methoxide or tetramethylammonium hydroxide as base. Because of the possible discriminating power of the non-aqueous solvents we can, for example, titrate a mixture of perchloric acid, hydrochloric acid, ethanoic acid, and phenol in a suitable solvent and obtain successive equivalence-points. Bases likewise could be distinguished in liquid ammonia. Liquid ammonia is of course just a little bit more inconvenient to manipulate.

7.8. POTENTIOMETRIC TITRATIONS

We have already discussed at some length the theory of pH meters and electrodes, and the characteristics of titration curves. In this section we put the theory into practice and discuss exactly how a potentiometric titration is done.

7.8.1. Theory of Potentiometric Titrations

In Section 7.2 we described the pH meter (and glass electrode) as a means of monitoring hydrogen-ion activity, and listed some advantages in its use compared with indicators.

Π Can you summarise four or five conditions necessary for a satisfactory indicator titration?

Amongst the possible answers you could give are:

- There has to be a region of sharp pH change with a small added volume of titrant.

- The pH range of the indicator has to lie within this pH change.

- The equivalence-point and end-point should be close and in this region of sharp pH change.

- The indicator volume should be minimal.

- The colour change should be clear and sharp.

- The sample should be colourless.

- The sample should not interfere with the indicator, eg if it contained Cl_2 gas, this might interfere.

- The sample must not be too dilute.

> You probably suggested that weak acid–weak base titrations cannot be done with indicators, (which is true), as they do not meet the first condition above.

Strong acid *vs* strong base, weak acid *vs* strong base and strong acid *vs* weak base are all usually considered for titration by use of an indicator, but especially for the last two the results are only approximate. As we have said, the first equivalence-point in, for example, titration of Na_2CO_3 *vs* HCl is rather poor.

To avoid most of these problems we use potentiometric titration in one form or another.

7.8.2. Method of Potentiometric Titration

Fig 7.8a shows the usual apparatus for potentiometric titration.

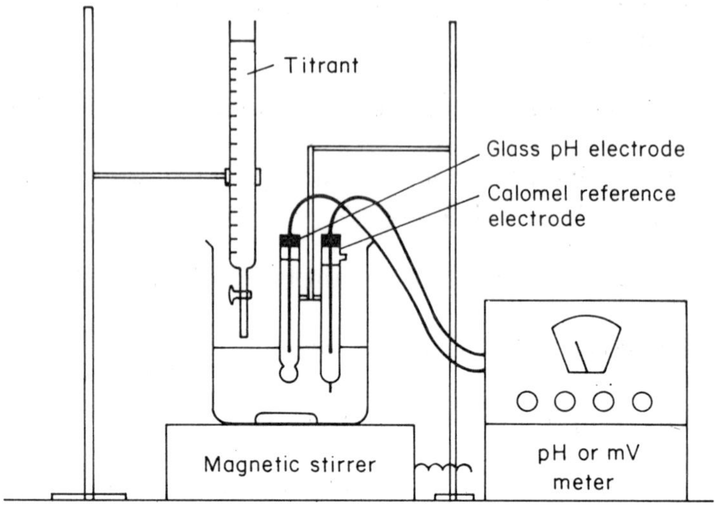

Fig. 7.8a. *Typical apparatus for a potentiometric titration*

The potential between the reference electrode and the indicator electrode is measured at the start and after the addition of small amounts of titrant (say each cm^3), and more closely near the equivalence-point, when readings start to change by larger amounts. After each addition, the solution is stirred well, usually magnetically, and the reading allowed to come to a steady value.

The normal response of a glass electrode is the Nernstian response, which at 25 °C is:

$$E = k + 0.0591 \text{ pH or } E = k - 0.0591 - \log a_{\mathrm{H}^+}$$

where k is the asymmetry (or junction) potential, an approximately constant factor for the individual glass electrode.

Whilst it would be possible to record the actual pH (or a_{H^+}) by calibration or by using a hydrogen electrode as indicator electrode, so called direct potentiometry, it is far more convenient to record the potential change with volume of titrant added. At 25 °C the usual change is 59.1 mV per unit pH change as long as there is no significant junction potential. Of course, as the pH changes sharply near the equivalence point there will be a sharp potential change in the same region.

Π What about the errors of the glass electrode; will these affect the potentiometric titration?

The asymmetry potential (and junction potentials) will show up as a constant or fairly constant bias throughout and will not significantly affect the equivalence-point response.

The acid and alkali errors of the glass electrode occur at extremes of pH, and so have little effect near the equivalence-point.

7.8.3. Calculation of Equivalence-point

Fig. 7.8b shows the result of a typical potentiometric titration.

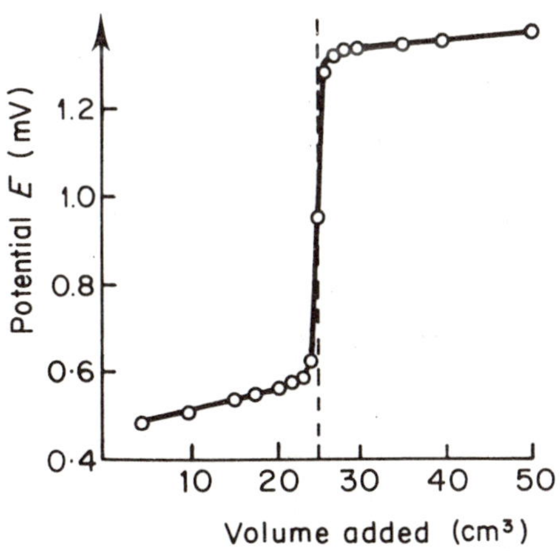

Fig. 7.8b. *Plot of potential E against volume added*

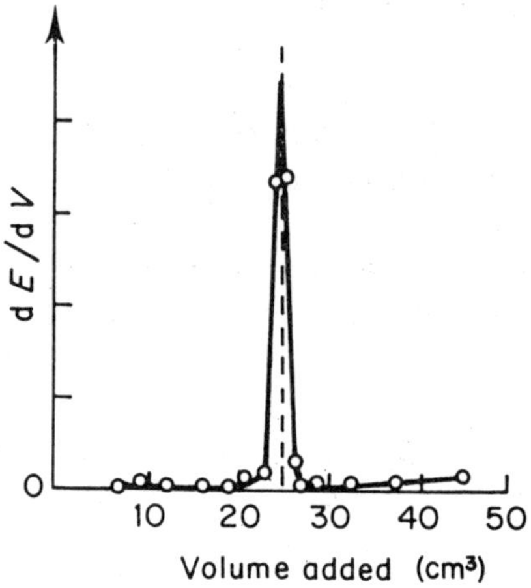

Fig. 7.8c. dE/dV *against volume added*

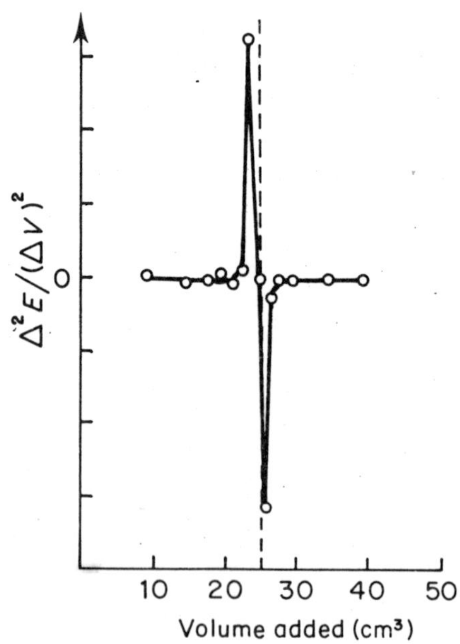

Fig. 7.8d. $\Delta^2 E/(\Delta V)^2$ *against volume added*

As we have discussed before, it is not difficult to determine the equivalence-point by using graphical aids to reduce the error in finding the point of inflection. An alternative is to use either $\mathrm{d}E/\mathrm{d}V$ or $\Delta^2 E/(\Delta V)^2$ (as in Fig. 7.8c and 7.8d respectively).

In Fig 7.8c, the first derivative indicates that equivalence would be at a maximum point, and possibly easier to determine. To plot it, intervals are taken (smaller intervals near the equivalence point) and $\Delta E/\Delta V$ is worked out and plotted against the mid-value of volume for that interval. Rarely does a reading give the maximum point and the method may not be superior to the first method, if only a few points near the equivalence point are plotted.

The same trouble applies to plotting the second derivative of the readings (as in Fig. 7.8d). Too few readings will make the construction of the vital part of the curve difficult. However we are looking for a change of slope from very positive to very negative, which is easier to find. The equivalence-point in most of these curves is at the zero value of $\Delta^2 E/\Delta V^2$ (as in Fig. 7.8d).

Π Why are errors intrinsically greater in the second-derivative curve?

In the first-derivative curve we take intervals of E and any error in the readings will be doubled in effect. In the second derivative we are taking intervals of intervals so the effect of errors should be greater. In fact the curve is less precise than the others, although the point of interest, the equivalence point is as easy to find.

7.8.4. Automated Methods

The sharp change in sign associated with the second derivative of potential with respect to volume is a very convenient signal for electronic detection. As the signal from the pH meter is a voltage, then if we keep the rate of supply of titrant constant, we should produce a signal in which potential varies with time. This can be differentiated once or twice electronically and the net signal can be recorded graphically, or fed to a microprocessor computer system for arithmetical processing to give the result on a VDU screen or similarly. The signal need not be passive but could be used as a feed-back signal operating the supply of titrant *via* hydraulics. We could thus arrange for slower and slower titration which would stop at the equivalence-point. The reading could then be taken in several ways.

Thus autotitrator systems can perform routine titrations in aqueous and non-aqueous media and, if auto-sampling is used, titrations on a large set of samples can be readily done. Considerable time can be saved once the initial setting up and testing of the equipment, and calibration with known samples has been done. The overall precision of such a method depends on the care taken with the setting up and proving, and rarely is better than that obtained by a skilled operator. It does, however, remove the tedium and cost of bulk titrations of similar samples. For short runs or unknown samples autotitrations may not be worthwhile. It must be remembered that a computer printout is only as reliable as the analytical input from all the equipment involved. For example, in a weak-acid–weak-base titration, a decision will have to be programmed in as to where to choose the end-point which may not, in fact, be possible. Also the titration rate and other parameters tend to be set, whereas a human operator would alter these as required by the particular titration.

7.9. APPLICATIONS

This brief section is intended to round up all we have done on neutralisation reactions. You should now be aware of the theoretical pitfalls in acid–base titrations, but at the same time I hope you realise that the techniques have much power. Some areas deserve more practical development: I believe non-aqueous titrations have been under-used in the past.

Finally we close this section with a list of routine assays that are still done by using the principles and practices of Part 7.

(*a*) Measurement of pH of effluents.
(*b*) Measurement of acid-base contents of gases (SO_2 NH_3, etc).
(*c*) Standardisation of acids and alkalis in commercial products.
(*d*) Acid-base content of products and raw materials.
(*e*) Acid values of fats, milk, cheese, etc.

(*f*) Alkalinity values of waters.
(*g*) Measurement of pH of blood and other biological fluids.
(*h*) Control of pH of water streams in industry.
(*i*) Use of buffers.
(*j*) Determination of purity of drugs, eg aspirin.
(*k*) Ammonia content of fertilisers.
(*l*) Kjeldahl nitrogen determination.

Objectives

At the end of Part 7 you should be able to:

- select and use primary standards correctly;
- select and use indicators for the neutralisation of simple acids and bases and of polyprotic acids and poly acid bases;
- describe the essential features and workings of a pH meter;
- select and use a range of buffer solutions;
- describe and perform titrations in non-aqueous solvents;
- perform potentiometric titrations;
- estimate and calculate the equivalence-point from indicator-based or potentiometric titration curves;
- design the best techniques for determining acids or bases in an industrial situation.

Self Assessment Questions and Responses

SAQ 1.1a

Which of the following represent the same equilibrium?

(*i*) $NH_{3(aq)} + H_2O_{(l)} \rightleftharpoons NH^+_{4(aq)} + OH^-_{(aq)}$

(*ii*) $NH_{3(aq)} + H_3O^+_{(aq)} \rightleftharpoons NH^+_{4(aq)} + H_2O_{(l)}$

(*iii*) $NH^+_{4(aq)} + OH^-_{(aq)} \rightleftharpoons NH_{3(aq)} + H_2O_{(l)}$

(*iv*) $H_2PO^-_{4(aq)} + H_2O_{(l)} \rightleftharpoons H_3O^+_{(aq)} + HPO^{2-}_{4(aq)}$

(*v*) $HPO^{2-}_{4(aq)} + H_3O^+_{(aq)} \rightleftharpoons H_2PO^-_{4(aq)} + H_2O_{(l)}$

(*vi*) $H_2PO^-_{4(aq)} + OH^-_{(aq)} \rightleftharpoons HPO^{2-}_{4(aq)} + H_2O_{(l)}$

Response

Careful examination shows that (*i*) and (*iii*), and (*iv*) and (*v*) are the same. The fact that one of the species appears in two similar looking equations is not sufficient for us to say that the equilibria are identical.

****

SAQ 1.1b

The expression below represents a general equilibrium system.

$$A + B \rightleftharpoons C + D$$

Mark the following statements as either true or false.

'The use of double arrows indicates that ...

(*i*) ... a mixture of reactants A and B reacts almost completely to give largely products, C and D.'

true ... false ...

(*ii*) ... a mixture of reactants C and D reacts almost completely to give largely products A and B.'

true ... false ...

(*iii*) ... if the reactants are A and B, then reaction takes place completely to give C and D, but if the starting materials are C and D, then complete reaction takes place to give A and B.'

true ... false ...

⟶

SAQ 1.1b (cont.)

(*iv*) ... the reaction takes place smoothly and rapidly.'

true ... false ...

(*v*) ... the reaction is reversible.'

true ... false ...

(*vi*) ... there are approximately equal amounts of A and B and C and D present when the reaction reaches equilibrium.'

true ... false ...

(*vii*) ... the system contains both A and B (some of which is undergoing reaction to C and D), and C and D (some of which is undergoing reaction to A and B).'

true ... false ...

(*viii*) ... the system can just as well be written as $C + D \rightleftharpoons A + B$.'

true ... false ...

(*ix*) ... the reaction is not stoichiometric.'

true ... false ...

(*x*) ... it is not possible to determine by experiment whether the starting materials were A + B or C + D.'

true ... false ...

Response

The correct sequence is as follows:

	true	false
(*i*)		√
(*ii*)		√
(*iii*)		√
(*iv*)		√
(*v*)	√	
(*vi*)		√
(*vii*)	√	
(*viii*)	√	
(*ix*)		√
(*x*)	√	

The double arrows indicate reversibility and consequently imply that we have an equilibrium system. Thus if (*v*) is true, (*vii*), (*viii*) and (*x*) follow logically.

Responses (*i*), (*ii*) and (*vi*) all imply some knowledge of the position of the equilibrium, ie some knowledge of the equilibrium constant, K_{eq}; as the question does not provide this, these assertions are false. The statement (*iii*) will be seen to be self-contradictory on close examination and is therefore false.

Statement (*iv*) refers to the *rate* of the reaction; the equilibrium expression does *not* convey information of this type.

The double arrows convey no information about stoichiometry and therefore (*ix*) is false

****

SAQ 1.1c

Several chemical systems are described below.

For each of these state whether the system is usefully described as an equilibrium system. For those processes which you identify as equilibria, write one or more equations which describe the system.

(*i*) Solid iodine in contact with iodine vapour in a closed container.

(*ii*) A sample of radio-labelled potassium iodide undergoing radioactive decay.

(*iii*) A solution of ethanoic acid (acetic acid) in water.

(*iv*) A sample of oxalic (ethanedioic) acid solution in an open flask being oxidised at 363 K (90 °C) according to the equation below.

$$2\,MnO_4^- + 5\,H_2C_2O_4 + 6\,H^+ \rightleftharpoons 2\,Mn^{2+} + 10\,CO_2 + 8\,H_2O$$

(*v*) A solution of ethanoic acid in 'heavy' water, D_2O.

(*vi*) A small volume of a dilute solution of iodine in tetrachloromethane (carbon tetrachloride) in contact with and enclosed by a larger volume of water, in an open flask.

(*vii*) A mixture of aqueous copper(II) sulphate and concentrated hydrochloric acid.

Response

(*i*) **Yes**, solid iodine in contact with iodine vapour in a closed container is an equilibrium system. It may be represented thus.

$$I_{2(s)} \rightleftharpoons I_{2(g)}$$

If the container was not closed, the system would not be in equilibrium in practical terms, as the gas-phase iodine would be effectively removed from the system and distributed to the atmosphere at large.

(*ii*) **No**, the disintegration of radioisotopes is not reversible.

(*iii*) **Yes**, there are in fact two equilibria of interest. One involves the dissociation of ethanoic acid and the other involves the self-ionisation of the solvent.

$$CH_3COOH + H_2O \rightleftharpoons CH_3COO^- + H^+$$

$$2\,H_2O \rightleftharpoons H_3O^+ + OH^-$$

If you answered **no** to this question you should re-read the section, and preferably consult a text such as D McQuarrie and P Rock, *General Chemistry*, Chapter 15, Freeman, 1984.

(*iv*) **No**. Under the conditions described in the question you can see that the carbon dioxide produced would be continuously removed from the system as a gas. Thus the system is not usefully described as an equilibrium system as it is 'open', and the reaction participates in this openness. However, if the system were closed so that carbon dioxide could not escape, it might be correctly described as an equilibrium system.

(*v*) **Yes**, the equilibria will be similar to those in part (*iii*).

However you should not overlook the range of other possibilities which will distribute protium (H) into the solvent and form deuterioethanoic acid, CH_3COOD.

eg

$$CH_3COOH + D_2O \rightleftharpoons CH_3COOD + HOD$$

$$CH_3COOH + D_2O \rightleftharpoons CH_3COO^- + HD_2O^+$$

$$CH_3COO^- + D_2O \rightleftharpoons CH_3COO^- + HD_2O^+$$

Remember that although we can write equations for these equilibria, this alone says nothing about the extent to which these processes occur or about the rates of reaction. The redistribution between ionisable protium (H) (or deuterium), and solvent protium (H) (or deuterium), is in fact quite rapid and is essentially statistically determined.

(*vi*) As this is an open system we should say that strictly it is not an equilibrium system. However, the loss of iodine to the atmosphere from the aqueous phase is negligibly small, so that in practice the process can be considered as an equilibrium process.

(*vii*) **Yes**, there are several equilibria here in addition to the self ionisation of water and the ionisation of HCl.

eg

$$Cu^{2+}(aq) + Cl^- \rightleftharpoons CuCl^+(aq)$$

$$Cu^{2+}(aq) + 4\,Cl^- \rightleftharpoons (CuCl_4)^{2-}(aq)$$

If you answered **no**, do not forget that metal ions in solution are solvated and that the equilibria are simply replacement of a solvent molecule by a chloride ion. Remember that the equilibria alone do not allow us to reach conclusions about the ionic composition of any particular mixture of copper sulphate and hydrochloric acid.

****

SAQ 1.2a

How is the value of K_{eq} for the equilibrium (*i*) $C + D \rightleftarrows A + B$ related to that for the equilibrium (*ii*) $A + B \rightleftarrows C + D$?

Response

$$K_{eq}\,(i) = \frac{[A][B]}{[C][D]} \text{ and } K_{eq}\,(ii) = \frac{[C][D]}{[A][B]}$$

You can see that one is the reciprocal of the other.

****

SAQ 1.2b

Indicate which of the following are correct completions of the statement:

'The law of mass action states that ...

(*i*) ... the rates at which equilibrium processes proceed are proportional to the indices which appear in the balanced equation.'

(*ii*) ... the active masses of reactants participating in a chemical reaction are always proportional to the product of the rate of reaction and the value of K_{eq}.'

(*iii*) ... the active masses of reactants participating in a chemical reaction are proportional to the rate constants for the forward and the reverse reaction.' ⟶

SAQ 1.2b (cont.)

> (*iv*) '... the rate of reaction is proportional to the active concentration of reacting substances each raised to the power of the index appearing in the balanced stoichiometric equation.'
>
> [The terms 'active mass', 'activity', and 'active concentration' may be regarded as synonymous].

Response

(*i*) This statement is part right and part wrong. The indices which appear in the balanced equation do indeed contribute to the expression for the rate, *but* the concentrations, or more correctly the active masses, also contribute to this.

(*ii*) This statement has been purposely garbled and is incorrect on two counts. First, the Law of Mass Action is concerned with the way in which other parameters are influenced *by* the active masses, *not vice-versa*. Secondly K_{eq} is the ratio of the rate constants for the forward and the back reaction at equilibrium.

(*iii*) This statement is misleading on two counts. As in (*ii*) above the active masses determine other things; if the system is at equilibrium we may be able to *compute* the active masses from some other parameter. Secondly, the rate constants or rate coefficients are by definition *constant* (and they will be different for the forward and the back reaction) hence the statement (*iii*) is not logically supportable.

(*iv*) This is the correct statement. Remember that in practice we use concentrations in place of active masses as long as the systems are at high dilution.

****

SAQ 1.2c

Derive expressions for K_{eq} for the reactions of stoichiometry (*i*) and (*ii*) below.

(*i*) $3A + B \rightleftharpoons 2C + 2D$

(*ii*) $A \rightleftharpoons 2B$

Response

$$K_{eq}(i) = \frac{[C]^2[D]^2}{[A]^3[B]}$$

$$K_{eq}(ii) = \frac{[B]^2}{[A]}$$

****

SAQ 1.2d

Write the equilibrium constant expression, K_{eq}, for the equilibria represented by the equations below.

For each example give also the units of K_{eq}.

(*i*) $PCl3 + P(OEt)3 \rightleftharpoons PCl2(OEt) + P(OEt)2Cl$

(*ii*) $N_2O_{4(g)} \rightleftharpoons 2\,NO_{2(g)}$

(*iii*) $HCOOH_{(aq)} + H_2O_{(l)} \rightleftharpoons H_3O^+_{(aq)} + HCOO^-{}_{(aq)}$

(*iv*) $N_{2(g)} + 3\,H_{2(g)} \rightleftharpoons 2\,NH_{3(g)}$

Response

$$(i)\ K_{eq} = \frac{[PCl_2(OEt)][P(OEt)_2Cl]}{[PCl_3]\ [P(OEt)_3]}$$

The numerator and the denominator have the same number of terms so K_{eq} is dimensionless.

$$(ii)\ K_{eq} = \frac{[NO_2]^2}{[N_2O_4]}$$

The concentration units do not cancel so the units of K_{eq} are mol dm^{-3}.

$$(iii)\ K_{eq} = \frac{[H_3O+][HCOO^-]}{[HCOOH][H_2O]}$$

As in (i) the concentration units cancel and K_{eq} is dimensionless.

$$(iv)\ K_{eq} = \frac{[NH_3]^2}{[N_2][H_2]^3}$$

This time we have the rather unusual unit of $mol^{-2}\ dm^6$.

This comes from

$$\frac{[mol\ dm^{-3}]^2}{[mol\ dm^{-3}][mol\ dm^{-3}]^3}$$

$$\equiv mol^{-2}\ dm^6$$

****

SAQ 1.2e

Explain why K_{eq} can never take negative values.

Response

The equilibrium constant is a *ratio* of quantities (including products of quantities) which can never have negative values. That is, one can never have negative concentrations or negative partial pressures, hence the ratio can never be negative.

****

SAQ 1.2f

The dissolution of silver sulphide in pure water involves the following equilibria.

$$Ag_2S_{(s)} \rightleftharpoons 2\,Ag^+ + S^{2-} \quad (1)$$

$$S^{2-} + H_2O \rightleftharpoons HS^- + OH^- \quad (2)$$

$$HS^- + H_2O \rightleftharpoons H_2S + OH^- \quad (3)$$

$$2\,H_2O \rightleftharpoons H_3O^+ + OH^- \quad (4)$$

Does the fact that the equilibrium constant for equation (1) is *ca.* 6×10^{-50} necessarily mean that silver sulphide will be insoluble in water?

What would be the effect of adding acid to the system?

Response

The very small value for K_{eq} would normally imply that the compound was of very low solubility. However in this case there are further reactions involving the anion. Therefore it does not necessarily follow from the data provided that silver sulphide will be insoluble.If this is to be true then values of K_{eq} for equations (2) and (3) must also be small. You will see from the equations that high values for K_{eq} for (2) and (3) would allow the conversion of the sulphur present into hydrogen sulphide, leaving the silver in solution as Ag^+ ions. This latter situation is not that observed by experiment.

Addition of acid to the system is predicted to lead to increased solubility of silver sulphide.If you did not reach this conclusion re-examine the equations with Le Chatelier's principle in mind.

Addition of acid leads to an increase in hydrogen-ion concentration and consequently reduces the hydroxide-ion concentration (equilibrium 4).Thus the processes represented by Eqs. (2) and (3) respond by moving to the right-hand side (as written in the question).In doing so they reduce the concentration of S^{2-}, and equilibrium (1) responds by allowing more silver sulphide to dissolve.

****

SAQ 1.2g

(*i*) If you are informed that the equilibrium constants for the reactions in water of ethylenediaminetetra-acetic acid, EDTA, with most divalent metals are above 10^{10}, does this mean that

(*a*) the complexes formed are very soluble,

(*b*) the equilibrium strongly favours the formation of the complex,

or

(*c*) the complexes are formed at pH values higher than 10? ⟶

SAQ 1.2g (cont.)

(*ii*) **What information is conveyed by the qualitative statement that the equilibrium constants for the dissociation of both aqueous ammonia and aqueous formic acid (methanoic acid) are moderately low?**

Response

(*i*) (*a*) Note that the question refers to the equilibrium constant for the reaction, it conveys no information about the solubility of the products.

(*b*) This is the correct interpretation; it follows from the fact that K_{eq} is the ratio of concentrations of the product(s) (here the complex) and the reactants.

(*c*) Although this assertion is in fact true for many complexes of EDTA, it does not follow directly from the information given.

(*ii*) Moderately low values of K_{eq} for aqueous formic acid and for aqueous ammonia indicate that they are both but weakly dissociated and therefore satisfactorily described in a qualitative sense as a weak acid and a weak base respectively. Note that in the question it was specified that the solutions were aqueous solutions. If these compounds are dissolved in other solvents different values of K_{eq} are obtained and consequently their acid/base behaviour can be vastly different.

****

SAQ 1.2h

Apply Le Chatelier's principle to the following.

(*i*) What will be the effect of an increase in temperature on the following equilibrium?

$$H_2 + I_2 \rightleftharpoons 2\,HI$$

$$\Delta H^{o} = +9.6 \text{ kJ mol}^{-1}$$

(*ii*) What will be the effect of an increase in pressure on the following equilibria?

(*a*) $SO_3 + NO \rightleftharpoons SO_2 + NO_2$

(*b*) $2\,NO_2 \rightleftharpoons 2\,NO + O_2$

(*iii*) A solution of hydrogen iodide, HI, in aerated water rapidly becomes browner as the temperature is raised. Is the reaction of HI with oxygen exothermic or endothermic?

(*iv*) Does the observation that the reaction of the oxalate ion, $(COO)_2^{2-}$, and the permanganate ion, MnO_4^-, proceeds more rapidly when the temperature is raised necessarily indicate that the reaction is endothermic?

Response

(*i*) The reaction is endothermic therefore you should expect the equilibrium to shift to the right hand side as the temperature increases.If you predicted the reverse, check that you understand the conventions for indicating exothermic (heat leaves the system) and endothermic (heat enters the system) reactions.

(*ii*) (*a*) There is no change in volume as the reaction proceeds therefore pressure alone will not affect the position of the equilibrium.

(*b*) As the reaction proceeds the volume change would be from 2 volumes to 3 volumes,therefore the equilibrium would shift to negate this, ie to move the position of equilibrium towards the left-hand side of the equation.

(*iii*) More iodine is produced as the temperature is raised therefore the reaction is endothermic. The equilibrium is shifting in the direction which will tend to negate the effect of the heat applied.

(*iv*) No, it does not necessarily follow; here the effect is principally one of increasing the rate of reaction.

If you answered 'yes' recall that it is the position of the equilibrium which enables us to draw conclusions about the heat change, *not* the rate at which the reaction proceeds.

****

SAQ 2.1a

Which of the following represent autoprotolysis?

(*i*) $HF + HF + \rightleftharpoons H_2F^+ + F^-$

(*ii*) $2\,H_2SO_4 \rightleftharpoons H_3SO_4^+ + HSO_4^-$

(*iii*) $2\,H_2SO_4 \rightleftharpoons H_2SO_3 + HSO_4^- + OH^-$

(*iv*) $HF + HF \rightleftharpoons HF_2^- + H^+$

(*v*) $2\,SO_2 \rightleftharpoons SO_3^{2-} + SO^{2+}$

(*vi*) $2\,HCN \rightleftharpoons H_2CN^+ + CN^-$

Response

(*i*), (*ii*) and (*vi*).

If you don't agree check that you have actually got an equation for proton transfer: (*iv*) is F-transfer rather than that of H^+, even though HF_2 is a known species; (*iii*) is nonsense as not only have we generated two negative charges without balancing positive charges but we have also incidentally reduced sulphur from S(+6) to S(+4) without a corresponding oxidation; (*v*) has no protons so there is no autoprotolysis. This reaction has however been proposed as an auto-ionisation reaction.

****

SAQ 2.1b

Mark the following as 'true' or 'false', within the context of the Brönsted–Lowry model.

		True	False
(*i*)	Electron donors are known as acids.		
(*ii*)	Acids react with proton donors.		
(*iii*)	Acids are themselves proton donors.		
(*iv*)	Water auto-ionises to produce H^+ therefore never acts as a base.		
(*v*)	Acids are proton acceptors.		
(*vi*)	Bases are proton acceptors.		

Response

Your responses should look like this;

	true	false
(*i*)		√
(*ii*)		√
(*iii*)	√	
(*iv*)		√
(*v*)		√
(*vi*)	√	

If you got any of these incorrect check through the text carefully. For (*iv*), water auto-ionises to produce *both* H^+ and OH^- and is therefore *capable* of acting either as an acid or as a base.

****

SAQ 2.1c

(1) Give the formula of the conjugate base of each of the following acids:

(*a*) C_6H_5COOH,

(*b*) HCO_3^-,

(*c*) H_2CO_3,

(*d*) NH_4^+.

(2) Give the formula of the conjugate acid of each of the following bases: ⟶

SAQ 2.1c (cont.)

(*a*) HPO_4^{2-},

(*b*) Cl^-,

(*c*) NH_3,

(*d*) OH^-.

Response

(1) (*a*) $C_6H_5COO^-$

(*b*) CO_3^{2-}

(*c*) HCO_3^-

(*d*) NH_3

(2) (*a*) $H_2PO_4^-$

(*b*) HCl

(*c*) NH_4^+

(*d*) H_2O

If you got any of these incorrect remember that an acid is a proton donor and a base a proton acceptor, therefore the acid always has more protons than its conjugate base.

loss of H^+

conjugate acid conjugate base

gain of H^+

****

SAQ 2.1d

(1) Which of the following statements are correct?

(*a*) H_3O^+ is the conjugate acid of H_2O.

(*b*) H_3O^+ is the conjugate acid of OH^-.

(*c*) H_3O^+ is the conjugate base of OH^-.

(*d*) OH^- is the conjugate base of H_3O^+.

(*e*) OH^- is the conjugate base of H_2O.

(*f*) H_2O is the conjugate base of H_3O^+.

(*g*) H_2O is the conjugate acid of OH^-.

(2) For the following equations arrange the materials in the form;

$$\text{acid}(1) + \text{base}(2) \rightleftharpoons \text{acid}(2) + \text{base}(1)$$

(*a*) $HCl + H_2O \rightleftharpoons H_3O^+ + Cl^-$

(*b*) $H_2O + CH_3CH_2COOH \rightleftharpoons H_3O^+ + CH_3CH_2COO^-$

(*c*) $H_2O + CO_3^{2-} \rightleftharpoons HCO_3^- + OH^-$

(*d*) $Al(H_2O)_4(OH)_2^+ + H_2O \rightleftharpoons H_3O^+ + Al(H_2O)_3(OH)_3$

(*e*) $CH_3COO^- + H_3O^+ \rightleftharpoons CH_3COOH + H_2O$

(*f*) $H_2O + Al(H_2O)_6^{3+} \rightleftharpoons AlOH(H_2O)_5^{2+} + H_3O^+$

(*g*) $CH_3NH_2 + H_2O \rightleftharpoons OH^- + CH_3NH_3^+$

Response

(1) The correct statements are (*a*), (*e*), (*f*) and (*g*). Statements (*b*), (*c*) and (*d*) involve the transfer of two protons and so are wrong.

(2) The equations should look like this:

$$\text{acid(1)} + \text{base(2)} \rightleftharpoons \text{acid(2)} + \text{base(1)}$$

(*a*) $$HCl + H_2O \rightleftharpoons H_3O^+ + Cl^-$$

(*b*) $$CH_3CH_2COOH + H_2O \rightleftharpoons H_3O^+ + CH_3CH_2COO^-$$

(*c*) $$H_2O + CO_3^{2-} \rightleftharpoons HCO_3^- + OH^-$$

(*d*) $$Al(H_2O)_4(OH)_2^+ + H_2O \rightleftharpoons H_3O^+ + Al(H_2O)_3(OH)_3$$

(*e*) $$H_3O^+ + CH_3COO^- \rightleftharpoons CH_3COOH + H_2O$$

(*f*) $$Al(H_2O)_6^{3+} + H_2O \rightleftharpoons H_3O^+ + Al(H_2O)_5OH^{2+}$$

(*g*) $$H_2O + CH_3NH_2 \rightleftharpoons CH_3NH_3^+ + OH^-$$

****

SAQ 2.2a

> Given that $K_w = 1 \times 10^{-14}$ mol^2 dm^{-6} at 25 °C, does this mean that the concentration of H^+ ions in pure water is (*i*) $10^{-14} M$, (*ii*) $10^{-7} M$, (*iii*) $10^{-28} M$, or (*iv*) $10^{+14} M$?

Response

K_w is the product of $[H^+]$ and $[OH^-]$. These ions are produced in *equal numbers* by the dissociation of water alone, hence $K_w = [H^+]^2$.

Thus $[H^+]$ is given by the square root of K_w, ie $10^{-7}M$.

****

SAQ 2.2b

The *International Critical Tables* give the following values for K_w at different temperatures. Units are all $mol^2 \, dm^{-6}$.

T °C	K_w
0	1.17×10^{-15}
50	5.49×10^{-14}
100	5.12×10^{-13}
150	2.34×10^{-12}
200	5.49×10^{-12}

Which of the following are true or false about pure water?

(*i*) $[H^+]$ increases and $[OH^-]$ decreases as the temperature increases.

(*ii*) $[H^+]$ decreases and $[OH^-]$ increases as the temperature increases.

(*iii*) Both $[H^+]$ and $[OH^-]$ increase as the temperature increases.

(*iv*) $[H^+]$ and $[OH^-]$ remain constant as the temperature increases.

(*v*) $[H^+]$ and $[OH^-]$ decrease as the temperature increases.

Response

The correct response is (*iii*). Remember that the dissociation of water must produce equal numbers of H^+ and OH^- in the absence of other compounds, hence (*i*) and (*ii*) which would alter these are not correct. K_w is in fact getting larger as the temperature increases, and as K_w is the product of two equal numbers, these numbers themselves must therefore also be increasing. Thus (*iv*) and (*v*) are also incorrect.

****

SAQ 2.4a	Look at the table of pK_a and pK_b values above, and remembering that the following classifications are rather broad, classify the compounds as strong, weak, or very weak electrolytes.

Response

w, w, s, s, vw, s, w, s, w, w, vw, vw and

w, w, vw, vw, vw, vw

If you got any of these wrong be careful to remember that pK_a = *minus* log K_a and pK_b = *minus* log K_b.

Thus K_a and pK_a 'move in opposite directions', higher K_a's lead to lower pK_a's. The same is true of course for K_b and pK_b.

****

SAQ 2.5a

Can you recall the relationship between pH and the hydrogen-ion concentration of a solution?

Try the following, assuming complete dissociation of the solute in water.

(*i*) Calculate the pH of 0.05*M*-HCl.

(*ii*) Calculate the pH of 0.05*M*-NaOH.

Response

$$pH = -\log[H^+]$$

(*i*) The molarity is given as 5×10^{-2}, therefore $[H^+] = 5 \times 10^{-2}$.

$$pH = -\log[H^+] = -\log(5 \times 10^{-2})$$

$$\therefore \quad pH = 1.3$$

(*ii*) Assuming complete dissociation $[OH^-] = [NaOH] = 5 \times 10^{-2}$

$$K_w = [OH^-][H^+]$$

$$[H^+] = K_w/[OH^-] = 10^{-14}/5 \times 10^{-2}$$

$$= 2 \times 10^{-13}$$

$$pH = -\log[H+] = -\log 2 \times 10^{-13}$$

$$pH = 12.7$$

****

SAQ 2.5b

> Calculate the hydrogen-ion concentration of the following materials for which pH values are given:
>
> lemon juice, 2.2; beer, 4.5; blood, 7.7; seawater, 8.3.

Response

Lemon juice, $6.31 \times 10^{-3} M$; beer, $3.16 \times 10^{-5} M$; blood, $1.99 \times 10^{-8} M$; seawater, $5.01 \times 10^{-9} M$.

A common problem here is forgetting to take account of the change in sign, eg if you got 158.4 for lemon juice, that is what you have done wrong. If you are using a calculator enter the pH, change its sign, then punch INV followed by LOG.

****

SAQ 2.5c

> Calculate the pH at each stage of a titration in which 14.0 cm^3 of $0.50M$-NaOH is added to 10.0 cm^3 of $0.50M$-HCl in 2.0 cm^3 portions. Do not forget the dilution which arises from the addition.

Response

The first calculation is to find the pH before the addition of any base.

As the acid is completely dissociated the concentration of hydrogen ions is obtained as in previous questions.

$$[H^+] = 0.50M$$

$$pH = -\log[H^+] = 0.301$$

During the additions we have:

$$[H^+] = \frac{(\text{moles acid} - \text{moles base})}{\text{new volume}}$$

After adding 2.0 cm^3,

$$[H^+] = \frac{(10.0 \times 0.50 - 2.0 \times 0.50)}{12.0} = 0.333\ M$$

$$pH = -\log 0.333 = 0.48$$

The remaining values up to 8.0 cm^3 are as follows:

Vol cm^3	4.0	6.0	8.0
$[H^+]$ M	0.214	0.125	0.055
pH	0.67	0.90	1.26

At the addition of 10.0 cm^3 of the base the acid is exactly neutralised so the hydrogen-ion concentration is given *via* K_W ($K_W = 10^{-14}$ $mol^2\ dm^{-6}$).

$$[H^+] = (K_W)^{\frac{1}{2}} = 10^{-7}\ mol\ dm^{-3}$$

$$pH = 7.0$$

After the end-point the hydrogen-ion concentration is given by computing the hydroxide ion concentration then working *via* K_W.

Vol added = 12.0 cm^3

$$OH^- = \frac{(12 \times 0.5 - 10 \times 0.5)}{22} = 0.045 \text{ mol dm}^{-3}$$

As $K_w = [H^+][OH^-], \quad [H^+] = K_w/[OH^-]$

$$[H^+] = 10^{-14}/0.045 = 2.22 \times 10^{-13} \text{ mol dm}^{-3}$$

$$\therefore \quad pH = 12.6$$

A similar calculation gives the pH after adding 14.0 cm^3 of base as 12.92.

****

SAQ 2.5d

By using a technique similar to that in the previous question, calculate the pH change when the following volumes of 0.01*M*-NaOH are added to 10.0 cm^3 of 0.005*M*-HCl: 0.0, 4.0, 4.5, 4.8, 5.0, 5.2, 5.5, 6.0 cm^3.

Response

The second part is calculated in an analogous manner,but note that this time the calculation gives you more points around the neutralisation region.

Vol/cm^3	0.0	4.0	4.5	4.8	4.9
$[H^+]/M$	0.00500	0.00071	0.00034	0.00014	0.000067
pH	2.30	3.15	3.46	3.87	4.17

At the neutralisation point the pH is 7.0 as before, the hydrogen-ion concentration is likewise obtained *via* K_w.

Vol/cm^3	5.2	5.5	6.0
$[OH^-]/(M)$	0.00013	0.00032	0.00062
$10^{11}[H^+]/(M)$	7.60	3.10	1.60
pH	10.12	10.51	10.80

****

SAQ 2.7a

Calculate the percentage error introduced by the simplifying assumption in the calculation of $[H^+]$ for the solutions below.

(*i*) $K_a = 10^{-4}$, $c = 0.010M$

(*ii*) $K_a = 10^{-4}$, $c = 0.001M$

(*iii*) $K_a = 10^{-2}$, $c = 0.010M$

(*iv*) $K_a = 10^{-2}$, $c = 0.001M$

Response

	method (*i*)	method (*ii*)	% error
(*i*)	1.00×10^{-3}	0.95×10^{-3}	5.3
(*ii*)	3.16×10^{-4}	2.70×10^{-4}	17
(*iii*)	1.00×10^{-2}	0.62×10^{-2}	61
(*iv*)	3.16×10^{-3}	0.92×10^{-3}	244

Case (*i*), as K_a gets larger, that is the proportion of dissociation gets larger we can see that the percentage error increases. This might be expected as the simplifying assumption is that the amount of dissociation is small enough to be neglected.

Case (*ii*), as the concentration gets smaller we can clearly see that the percentage error increases.

$**********************************$

SAQ 2.7b

For the dissociation of HCN, $K_a = 4.93 \times 10^{-10}$. Calculate the pH of 0.10*M* HCN in water. What is the effect of making the solution (*a*) 1.00*M* in sodium cyanide, (*b*) 1.00*M* in HCl?

Response

pH of the original solution = 5.1

pH when made molar in NaCN = 10.3

Effect of molar HCl is to lower the CN^- ion concentration to $4.3 \times 10^{-11}M$.

If you do not agree with these figures first check the arithmetic; if there is no simple arithmetical error examine the general strategy of attacking the problem. A good sequence is to work from the dissociation equation to the expression for K_a (or in other examples you might want K_b etc).Then rearrange the expression to isolate the term you want before attempting to put in any numerical values. After isolating the term you want you should be able to detect whether you have one or more unknowns. If there is more than one unknown you must examine the stoichiometry to obtain permissible assumptions or equalities.

$$HCN \rightleftharpoons H^+ + CN^-$$

$$K_a = \frac{[H^+][CN^-]}{[HCN]}$$

(*i*) initially $[H^+] = [CN^-]$

therefore $[H^+]^2 = 4.93 \times 10^{-10} \times 0.10$

$$[H^+] = 7.02 \times 10^{-6}$$

$$pH = 5.15$$

(*ii*) $$[H^+] = \frac{K_a \times [HCN]}{[CN^-]}$$

$$[H^+] = 4.93 \times 10^{-11}$$

$$pH = 10.3$$

(*iii*) this time $[H^+]$ is 1.00.

$$[CN^-] = \frac{K_a \times [HCN]}{[H^+]}$$

$$[CN^-] = 4.93 \times 10^{-11}$$

Note that the addition of the strong acid HCl has severely depressed the dissociation of HCN.

****

SAQ 2.10a

> **Classify the following salts as 'acid', 'basic' or 'neutral' according to their behaviour in water:**
>
> **KCN, $NaNO_3$, NaH_2PO_4, $Zn(ClO_4)_2$**
>
> **Give the relevant equations when they are not neutral.**

Response

The ions can be classified as follows, thus permitting the classification shown for the salts.

	n b	n n	n a	a n
	KCN	$NaNO_3$	NaH_2PO_4	$Zn(ClO_4)_2$
a			√	√
b	√			
n		√		

$$CN^- + H_2O \rightleftharpoons HCN + OH^-$$

$$H_2PO_4^- + H_2O \rightleftharpoons HPO_4^{2-} + H_3O^+$$

$$Zn^{2+} + 2H_2O \rightleftharpoons Zn(OH)^+ + H_3O^+$$

****

SAQ 2.10b

> **Use reasoning similar to that above to derive an expression relating K_h for a salt of a strong base and weak acid to K_w and K_a.**

Response

$$A^- + H_2O \rightleftharpoons HA + OH^- \quad K_{eq} = \frac{[HA][OH^-]}{[A^-][H_2O]}$$

$$K_h = \frac{[HA][OH^-]}{[A^-]} = \frac{[HA][OH^-][H^+]}{[A^-][H^+]} = K_w/K_a$$

****

SAQ 2.10c

By using reasoning similar to that above, show that in the hydrolysis of a salt of a weak acid and a strong base, pH is given by:

$\frac{1}{2}\text{p}K_w + \frac{1}{2}\text{p}K_a + \frac{1}{2}\log c.$

Response

For concentration, c, and noting that $[OH^-] = [HA]$

$$K_h = \frac{K_w}{K_a} = \frac{[HA][OH^-]}{[A^-]} = \frac{[OH^-]^2}{c}$$

Also $[OH^-] = K_w/[H^+]$

$\therefore \quad [OH^-]^2 = K_w/K_a = K_w^2/[H^+]^2$

$\therefore \quad \dfrac{1}{[H^+]} = (cK_aK_w)^{\frac{1}{2}}$

$\text{pH} = \frac{1}{2}\log c + \frac{1}{2}\text{p}K_a + \frac{1}{2}\text{p}K_w$

****

SAQ 3.2a

Given that the solubility product of copper(I) chloride at 25 °C is $1.20 \times 10^{-6} M^2$, calculate the solubility of copper(I) chloride at equilibrium in water at 25 °C.

If you had a sample which originally contained 1.0 g CuCl in contact with one dm^3 of water at 25 °C, what would be the approximate percentage error introduced by assuming that copper(I) chloride was completely insoluble?

Response

We are given $K_{sp} = 1.20 \times 10^{-6}$, thus

$$[Cu^+][Cl^-] = 1.20 \times 10^{-6}$$

$\therefore$ the solubility S is given by $K_{sp}^{\frac{1}{2}}$ as,

$$S = (1.20 \times 10^{-6})^{\frac{1}{2}}$$

$$= 1.09 \times 10^{-3} \text{ mole dm}^{-3}$$

Relative atomic and molecular masses are Cu, 63.5 and Cl,35.5, thus CuCl = 99.0.

$$\therefore \quad S = 1.09 \times 99.0 \times 10^{-3} \text{ g dm}^{-3}$$

$$= 108 \text{ mg dm}^{-3}$$

Therefore for a sample originally containing one gram in 1.0 dm^{-3} of water, *ca* 10% would be lost to solution.

****

SAQ 3.2b

> Use a treatment similar to that above to obtain an expression for the solubility of a compound MX_3 in terms of K_{sp}.

Response

Here $[X^-] = 3 \times [M^{3+}]$ and $S = [M^{3+}]$ as before.

Therefore $S = (K_{sp}/27)^{\frac{1}{4}}$

****

SAQ 3.3a

> Given that $K_{sp}(AgCl) = 1.80 \times 10^{-10}$ mol^2 dm^{-3} at 25 °C, calculate the solubility of silver chloride in the following solutions:
>
> (*i*) pure water,
> (*ii*) 0.0010*M*-KCl,
> (*iii*) 0.0100*M*-KCl,
> (*iv*) 0.1000*M*-KCl.

Response

(*i*) $K_{sp} = [Ag^+][Cl^-] = 1.8 \times 10^{-10}$

$\therefore \quad S = [Ag^+] = (K_{sp})^{\frac{1}{2}} = 1.34 \times 10^{-5}$ mol dm^{-3}

(*ii*) Here we assume that the concentration of chloride ion from the dissolved silver chloride is negligible in comparison with that from added potassium chloride.

In $0.0010M$-KCl $[Cl^-] = 1.00 \times 10^{-3}M$

$$K_{sp} = [Ag^+][Cl^-] = [Ag^+][10^{-3}]$$

$$[Ag^+] = S = K_{sp}/10^{-3} = 1.80 \times 10^{-7} \text{ mol dm}^{-3}$$

Similar working to the above gives for

(*iii*) $S = 1.80 \times 10^{-8}$ mol dm^{-3}, and

(*iv*) $S = 1.80 \times 10^{-9}$ mol dm^{-3}.

****

SAQ 3.3b

What is the molar concentration of dissolved barium chromate in a solution which is made 0.060 M in barium chloride? ($K_{sp} = 1.2 \times 10^{-10}M^2$).

Response

As K_{sp} is very small we initially assume that numbers of barium ions from the barium chloride far exceed those from the barium chromate, ie $[Ba^{2+}] = 0.06M$.

$$K_{sp} = [Ba^{2+}][CrO_4^{2-}] = 1.2 \times 10^{-10}M^2$$

$$\therefore \quad [CrO_4^{2-}] = K_{sp}/[Ba^{2+}]$$

$$= \frac{1.2 \times 10^{-10}M^2}{0.6 \times 10^{-2}M}$$

$$= 2 \times 10^{-8}M.$$

As one mole of barium chromate produces one mole of chromate ion the concentration of barium chromate is also $2 \times 10^{-8} M$.

****

SAQ 4.1a

> **Decide whether each of the following statements is true or false.**
>
> **(*i*) When a coordination compound is formed the ligand is behaving as a Lewis acid**
>
> **T / F**
>
> **(*ii*) The equilibria involved in the formation of complex species usually involve the displacement of one donor group by another.**
>
> **T / F**
>
> **(*iii*) Ligand molecules always contain neutral donor groups.**
>
> **T / F**

Response

Your replies should look like this:

(*i*) False. The ligand is the donor group involved in coordination *to* the positively charged metal ion, ie the ligand is a Lewis base and it is the metal ion which is behaving as a Lewis acid.

(*ii*) True. This is usually true and it is tempting to say that it is always true, but there are examples where mechanistic studies indicate coordination of extra ligands without displacement reactions.

(*iii*) False. The requirement of the ligand is that it should be a *donor*, thus ligands may be neutral eg NH_3 or charged eg Br^-, or even contain both types of donor site eg $NH_2CH_2COO^-$. It would not be expected, of course, that positively charged species would act as ligands to positively charged metal ions. There are special cases where species such as NO^+ can act as ligands with zero- valent metal species, but these involve special bonding mechanisms and are not important to our principal interest in analytical chemistry.

****

SAQ 4.1b

The addition of ammonia to $CoCl_3$ gives rise to the compound below. Identify the components by marking the grid as appropriate.

$Co(NH_3)_6Cl_3$ or

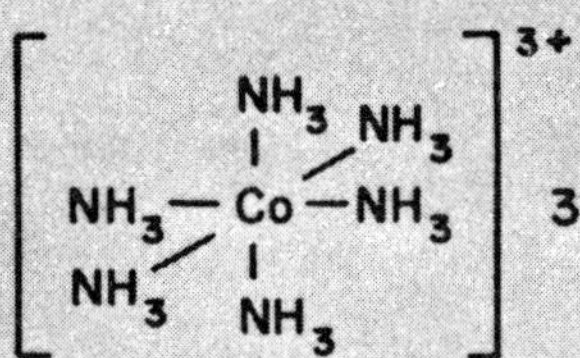

$3Cl^-$

Cl^- NH_3 Co^{3+} $Co(NH_3)_6^{3+}$ $Co(NH_3)_6Cl_3$

ligand,

complex-ion,

nucleophile,

coordination
-compound,

Lewis acid,

Lewis base.

Response

Your grid should look like this:

	Cl^-	NH_3	Co^{3+}	$Co(NH_3)_6^{3+}$	$Co(NH_3)_6Cl_3$
ligand,		√			
complex-ion,				√	
nucleophile,		√			
coordination -compound,					√
Lewis acid,			√		
Lewis base,		√			

If you do not agree with these re-read the text carefully.

****

SAQ 4.2a

> **For the equilibrium involving Ni^{2+} and NH_3 write out the expressions for the following, omitting any participation by water molecules:**
>
> **K_1, β_2, K_3, and β_6.**

Response

$$K_1 = \frac{[Ni(NH_3)]}{[Ni^{2+}][NH_3]}$$

$$\beta_2 = \frac{[Ni(NH_3)_2]}{[Ni^{2+}][NH_3]^2}$$

$$K_3 = \frac{[Ni(NH_3)_3]}{[Ni(NH_3)_2][NH_3]}$$

$$\beta_6 = \frac{[Ni(NH_3)_6]}{[Ni^{2+}][NH_3]^6}$$

If your responses do not agree with those given check that you are distinguishing the stepwise stability constants (K) from the overall or total stability constants (β). Remember that for the stepwise equilibrium one is starting from the previous complex.

****

SAQ 4.2b

Which of the following is the correct concluding phrase to the phrase below?

'In analytical procedures using unidentate ligands it is common to require large excesses of the ligand because ... '

(*i*) ... large excesses of all reagents are good laboratory practice.'

(*ii*) ... one always needs large excesses to form any coordination compound.'

(*iii*) ... large excesses are frequently needed to ensure the predominance of only one complex species.'

(*iv*) ... stability constants are always small for coordination compounds.'

Response

The correct response is (*iii*).

(*i*) is not true for general laboratory practice and unnecessary excesses can at times cause problems as well as being expensive in reagents.

(*ii*) and (*iv*) are certainly not true. Many coordination compounds have quite high stability constants and do not require large excesses. This is of course essential if coordination compounds are to be used in titrations.

****

SAQ 4.2c

For the hypothetical series of complex equilibria involving stepwise reaction of a ligand L with a metal M in the molar concentration ratio L:M of 4:1, which of the following descriptive phrases would be expected to match the data?

	A	B	C	D
K_1 =	10^6	10^3	10^3	10^5
K_2 =	$10^{0.7}$	$10^{2.8}$	10^4	$10^{4.5}$
K_3 =	$10^{0.5}$	$10^{0.1}$	$10^{4.5}$	10^4
K_4 =	$10^{0.1}$	$10^{0.02}$	10^6	$10^{3.5}$

(*i*) There should be a mixture with significant concentrations of all species, ML, ML_2, ML_3, and ML_4 along with unbound L.

(*ii*) The species ML_4 will predominate.

(*iii*) The mixture will be largely ML and ML_2 with some unbound L.

(*iv*) There will be a significant amount of unbound L and the principal complex will be ML.

Response

The match is as follows

(*i*) ... D

(*ii*) ... C

(*iii*) ... B

(*iv*) ... A

****

SAQ 4.3a

Which of the following ligands would you expect to form chelates?

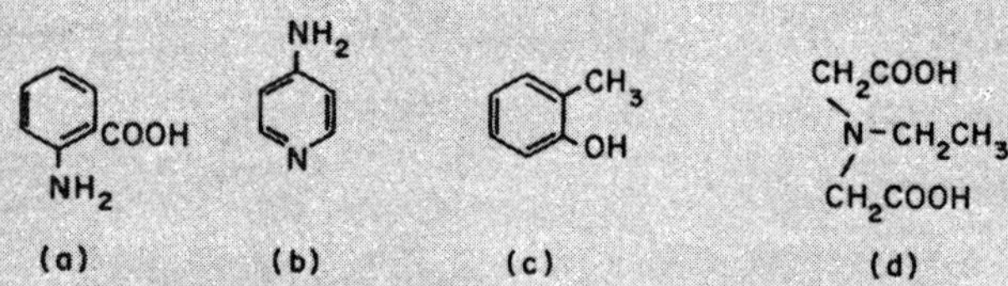

Response

(*a*) and (*d*) In both cases ionisation of the carboxy group is necessary for chelation.

(*b*) has N atoms which are not well placed for forming 5-, or 6-membered rings.

(*c*) has only one donor group.

****

SAQ 4.3b

> **Which of the following is most correct?**
>
> **The formation constants of chelates are ...**
>
> **(*i*) ... usually greater ...**
>
> **(*ii*) ... marginally greater ...**
>
> **(*iii*) ... always much greater ...**
>
> **... than the formation constants for closely related non-chelated systems.**

Response

The correct response is (*i*) but the inclusion of 'usually' is important as in some cases steric features can destroy the additional stability generally associated with chelate formation. Thus (*ii*) is occasionally true and (*iii*) is also occasionally true if the word 'always' is removed.

****

SAQ 4.4a

Which of the curves in the figure would you expect to correspond in general form to that for the addition of the bidentate ligand 1,10-phenanthroline to a solution of a metal ion with a coordination maximum of four?

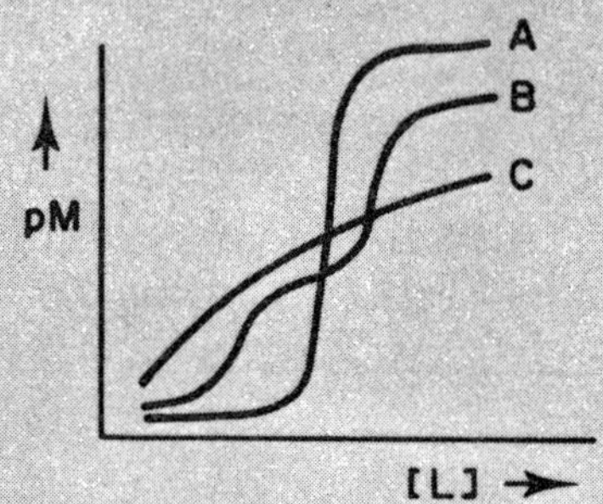

Response

The correct curve is B; A is the curve expected for a quadridentate ligand, and C is that for four unidentate ligands.

****

SAQ 5.1a

What is the oxidation state of the following?

(*i*) Mn in MnO_4^-,

(*ii*) Cu in $[Cu(NH_3)_4]^{2+}$,

(*iii*) Se in SeO_3^{2-},

(*iv*) Cr in $Cr_2O_7^-$, →

SAQ 5.1a (cont.)

(*v*) **I in I_2,**

(*vi*) **I in HI,**

(*vii*) **N in N_2H_4.**

Response

(*i*) +7

(*ii*) +2

(*iii*) +4

(*iv*) +6

(*v*) 0

(*vi*) −1

(*vii*) −2

If you found these difficult you should consult an introductory text such as D McQuarrie and P Rock, *General Chemistry*, Chapter 20, Freeman, 1984, or W Masterson, E J Slovoinski, and Stanitski *Chemical Principles*, Chapter 24, Saunders, New York, 1985; alternatively almost any introductory inorganic text will deal with this topic.

****

SAQ 5.1b

> **Which is the reactant undergoing *oxidation* in the following equilibria?**
>
> (*i*) $SO_3^{2-} + I_2 + H_2O \rightleftharpoons SO_4^{2-} + 2H^+ + 2I^-$
>
> (*ii*) $12H^+ + 4MnO_4^- + 5Sb_2O_3 \rightleftharpoons 4Mn^{2+} + 6H_2O + 5Sb_2O_5$
>
> (*iii*) $2Cu^{2+} + 4I^- \rightleftharpoons 2CuI + I_2$
>
> (*iv*) $2MnO_4^- + 10I^- + 16H^+ \rightleftharpoons 2Mn^{2+} + 5I_2 + 8H_2O$

Response

(*i*) O_3^{2-} ... is oxidised to SO_4^{2-}

(*ii*) Sb_2O_3 ... is oxidised to Sb_2O_5

(*iii*) I^- ... is oxidised to I_2

(*iv*) I^- ... is oxidised to I_2

If you got any of these wrong have another look at the oxidation numbers of the 'central' elements (S, I, Mn, Sb, and Cu). Note that in some cases this goes up (the species undergoing oxidation), in others down (the species undergoing reduction), eg for (*i*) S(+4) is oxidised to S(+6) and I(0) is reduced to I(−1).

****

SAQ 5.1c

For the following equilibria which reactant is the oxidant?

(*i*) $10\,Cl^- + 2\,BrO_3^- + 12\,H^+ \rightleftharpoons 5\,Cl_2 + Br_2 + 6\,H_2O$

(*ii*) $2\,S_2O_3^{2-} + I_2 \rightleftharpoons S_4O_6^{2-} + 2\,I^-$

(*iii*) $5\,V^{2+} + 3\,MnO_4^- + 24\,H^+ \rightleftharpoons 5\,V^{5+} + 3\,Mn^{2+} + 12\,H_2O$

(*iv*) $Fe^{3+} + Ti^{3+} \rightleftharpoons Fe^{2+} + Ti^{4+}$

Response

(*i*) BrO_3^- ... oxidises Cl^- to Cl_2

(*ii*) I_2 ... oxidises $S_2O_3^{2-}$ to $S_4O_6^{2-}$

(*iii*) MnO_4^- ... oxidises V^{2+} to V^{5+}

(*iv*) Fe^{3+} ... oxidises Ti^{3+} to Ti^{4+}

If you were not clear about any of these, again decide on the oxidation number of the species involved. The oxidant is the substance for which the oxidation number of the 'central' species decreases.

****

SAQ 5.1d

Write balanced equations for the half-reactions which represent the changes below and use them to produce balanced equations for the reactions which follow (*i*) to (*v*).

$$ClO_3^- \rightarrow Cl^-$$

$$RNO_2 \rightarrow RNH_2$$

$$S^{2-} \rightarrow SO_4^-$$

$$I^- \rightarrow I_2$$

$$Ti^{3+} \rightarrow Ti^{4+}$$

$$OBr^- \rightarrow Br^-$$

$$H_2O_2 \rightarrow H_2O$$

$$IO_3^- \rightarrow I_2$$

$$As_2O_3 \rightarrow As_2O_5$$

(*i*) The reaction of chlorate ions and iodide ions in acidic media.

(*ii*) Reduction of nitro compounds by using titanium(+3).

(*iii*) The reaction of sulphide ions with bromate(I) ions (hypobromite).

(*iv*) The oxidation of arsenic(III) oxide by hydrogen peroxide.

(*v*) The reaction of iodate ions with iodide ions.

Response

$ClO_3^- + 6H^+ + 6e = Cl^- + 3H_2O$

$RNO_2 + 6H^+ + 6e = RNH_2 + 2H_2O$

$S^{2-} + 4H_2O = SO_4^{2-} + 8H^+ + 8e$

$2I^- = I_2 + 2e$

$Ti^{3+} = Ti^{4+} + e$

$OBr^- + 2H^+ + 2e = Br^- + H_2O$

$H_2O_2 + 2H^+ + 2e = 2H_2O$

$2IO_3^- + 12H^+ + 10e = I_2 + 6H_2O$

$As_2O_3 + 2H_2O = As_2O_5 + 4H^+ + 4e$

If you got any of these wrong, please ensure that you do the balancing stages in the right order. That is, balance oxygen atoms by adding water, *then* balance hydrogen atoms by adding H^+, *then* add electrons to balance the charge. Do not try to balance the charges before you have the correct atom balance.

Your final equations should be as below: remember that you must adjust the half-equations so that the electrons cancel.

(*i*) $ClO_3^- + 6H^+ + 6I^- = Cl^- + 3I_2 + 3H_2O$

(*ii*) $RNO_2 + 6Ti^{3+} + 6H^+ = RNH_2 + 6Ti^{4+} + 2H_2O$

(*iii*) $S^{2-} + 4OBr^- = SO_4^{2-} + 4Br^-$

(*iv*) $As_2O_3 + 2H_2O_2 = As_2O_5 + 2H_2O$

****

SAQ 5.1e

Mark the following statements about redox reactions as 'true' (T) or 'false' (F).

(*i*) Redox reactions are either oxidising or reducing reactions depending on the prevailing conditions.

(T) (F)

(*ii*) If the oxidant is added to the other reagent the process is always described as an oxidation and reduction does not take place.

(T) (F)

(*iii*) Redox reactions involve both a reduction and an oxidation process occurring simultaneously.

(T) (F)

(*iv*) In redox processes the oxidation number of the oxidant increases.

(T) (F)

(*v*) In redox processes the electrons are transferred to the reductant from the oxidant.

(T) (F)

(*vi*) The change in oxidation numbers of oxidant and reductant arises from electron transfer.

(T) (F)

(*vii*) The oxidising agent is the material which gains electrons during a redox process.

(T) (F)

⟶

SAQ 5.1e (cont.)

(*viii*) The change in the oxidation number of the oxidant must be equal to that of the reductant.

(T) (F)

(*ix*) Strong oxidising agents generally contain central elements with high positive oxidation numbers.

(T) (F)

Response

Your responses should look like this:

	true	false
(*i*)		√
(*ii*)		√
(*iii*)	√	
(*iv*)		√
(*v*)		√
(*vi*)	√	
(*vii*)	√	
(*viii*)		√
(*ix*)	√	

(*i*), (*ii*), and (*iii*). Although chemists often talk about oxidising one reagent with another, for example A is oxidised by B, it is just as true to say that B is reduced by A. Changing the conditions may change the kinetics and may displace the equilibrium but the 'amount of oxidation' taking placed under set conditions must be balanced by the 'amount of reduction' in a redox reaction. Hence (*i*) & (*ii*) are false, (*iii*) is therefore true.

(*iv*), (*v*), (*vi*) and (*vii*). The oxidant starts with a high positive oxidation number and receives electrons from the reductant as the reaction proceeds, hence the oxidation number of the oxidant decreases while that of the reductant increases (or becomes less negative). Clearly (*iv*) and (*v*) are the reverse of this and are therefore false while (*vi*) and (*vii*) are true.

(*viii*) This does not *necessarily* follow as one mole of an oxidant may react with more than (or less than) one mole of the reductant. The total number of electrons transferred is of course the same, but because of stoichiometry oxidation number changes may be different.

(*ix*) The high positive oxidation number is the result of the removal of electrons. Hence this enables species to accept electrons and consequently lower the oxidation number, ie a high positive oxidation number characterises materials normally called *strong* oxidising agents.

****

SAQ 5.1f

By using the data in Fig. 5.1c (E^{o} values), state whether the following statements are true (T) or false (F), when the constituents are at unit activity.

(*i*) Fe^{3+} is expected to oxidise Br^{-}.

(T) (F)

(*ii*) MnO_4^- is a more powerful oxidant than $Cr_2O_7^{2-}$.

(T) (F)

(*iii*) IO_3^- is capable of oxidising chromium(+2) to chromium(+3) but not to chromium(+6).

(T) (F)

⟶

SAQ 5.1f (cont.)

(*iv*) Metallic zinc should be insoluble in solutions containing ferric ions. (T) (F)

(*v*) Potassium bromate will not be oxidised by potassium dichromate. (T) (F)

(*vi*) Cadmium metal placed in a solution of zinc ions should lead to the plating-out of zinc metal. (T) (F)

(*vii*) Although tin(+2) is properly described as usually having reducing properties it will not reduce bromine to bromide. (T) (F)

Response

Your responses should be as follows;

	true	false
(*i*)		√
(*ii*)	√	
(*iii*)	√	
(*iv*)		√
(*v*)	√	
(*vi*)		√
(*vii*)		√

If you don't agree with any of these, check carefully to see that you are considering the correct half-equation. Remember that the half-reaction with the more positive value of E^0 will proceed as a reduction, ie *a component in it gets reduced, or alternatively this does the oxidising*.

****

SAQ 5.1g

For an electrochemical cell in which the cell reaction is;

$$Zn^{o}_{(s)} + Cu^{2+} \rightleftharpoons Cu^{o}_{(s)} + Zn^{2+}$$

predict:

(*i*) the effect of increasing the concentration of Cu^{2+} ions on the cell voltage,

(*ii*) the effect of increasing the concentration of Zn^{2+} ions on the cell voltage,

(*iii*) the effect of increasing the size of the zinc electrode on the cell voltage.

Response

(*i*) and (*ii*)

Le Chatelier's principle allows us to make a qualitative assessment of these effects. In the absence of values for E^{o} and data on actual concentrations a quantitative evaluation is not possible.

Following the Le Chatelier's reasoning we note that an increase in the concentration of copper(+2) ions will lead the system to offset this by shifting the equilibrium to the right hand side. In other words there is an increase in the driving force, and E_{cell} increases.

Notice that zinc ions are a product of the reaction, so we expect the converse. An increase in zinc-ion concentration pushes the equilibrium position over to the left-hand side, ie the driving force decreases and E_{cell} is expected to decrease.

Let us now follow this reasoning through by using the Nernst equation (note that this is a 2-electron change, so n in the equation is 2).

For the copper half-reaction;

$$E_{Cu} = E_{Cu^{o}} - \frac{0.0591}{2} \log \frac{[Cu_{(s)}]}{[Cu^{2+}]}$$

for the zinc half-reaction;

$$E_{Zn} = E^{o}_{Zn} - \frac{0.0591}{2} \log \frac{Zn_{(s)}}{[Zn^{2+}]}$$

Now $E_{cell} = E_{Cu} - E_{Zn}$;

(*i*) If Cu^{2+} increases the second term in the equation decreases hence E_{Cu} increases and E_{cell} increases.

(*ii*) If Zn^{2+} increases, by similar reasoning E_{Zn} increases, thus E_{cell} decreases.

If you experienced difficulty with these questions be careful to check that you make proper allowance for the negative signs and slowly follow the reasoning through.

(*iii*) Even though a term in $[Zn_{(s)}]$ appears in the equation, we recall that solids have unit activity irrespective of the quantity present, hence the size of the zinc electrode does not effect the value of the cell potential.

****

SAQ 5.1h

For an electrochemical cell in which the cell reaction is:

$$H_{2(g)} + 2\,AgCl_{(s)} \rightleftharpoons 2\,Ag_{(s)} + 2\,H^+_{(aq)} + 2\,Cl^-_{(aq)}$$

what is the effect of the following factors on the cell potential?

(*i*) Decreasing the pH,

(*ii*) Increasing the pressure of H_2,

(*iii*) Decreasing the amount of silver chloride present,

(*iv*) Increasing the molarity with respect to chloride ions.

Try making the prediction first by using the Nernst equation, then see whether Le Chatelier type reasoning leads to the same conclusion.

Response

(*i*) decrease,

(*ii*) increase,

(*iii*) no effect,

(*iv*) decrease.

If you disagree with these answers, approach the problem by writing out the equations for the half-reactions, then the equations for E in each case.

Remember that $E_{cell} = E_{AgCl/Ag} - E_{H^+/H_2}$

You should find that Le Chatelier type reasoning gives the same results.

****

SAQ 5.2a

Calculate the potential for the half-cell consisting of 0.0100*M* KBr in the presence of liquid bromine. (In practice a platinum electrode is used for this system to avoid complications due to oxidation of other electrode materials). Note that in this example the aqueous phase is always saturated with bromine.

$$Br_{2(l)} + 2e \rightleftharpoons 2\,Br^- \qquad E^\circ = 1.065\ V$$

Response

The potential is 1.183 V.

If you don't agree check that you have not slipped up with signs and remember that as liquid bromine is always present its activity is unity.

From the Nernst equation:

$$E = 1.065 - \frac{0.0591}{2}\log\frac{[Br^-]^2}{[Br_2]}$$

Hence
$$E = 1.065 - \frac{0.0591}{2}\log\frac{[Br^-]^2}{[1.0]}$$

$$E = 1.065 - \frac{0.0591}{2}(-4.0)$$

$$= 1.183 \text{ V.}$$

****

SAQ 5.2b

Calculate the equilibrium constant for the reaction between iron(+2) and manganate(+7) (the permanganate ion).

MnO_4^-/Mn^{2+}, $E^o = +1.51$ V.

Fe^{3+}/Fe^{2+}, $E^o = +0.78$ V.

Response

$$K_{eq} = 5.75 \times 10^{61}$$

We will work through this stage by stage.

Stage (*a*) ... write the balanced equations for the half-reactions then the equation for the reaction as a whole.

$$MnO_4^- + 8H^+ + 5e \rightleftharpoons Mn^{2+} + 4H_2O, \quad E^o = 1.51$$

$$Fe^{3+} + e \rightleftharpoons Fe^{2+}, \quad E^o = 0.78$$

The values for the standard electrode potentials indicate that the manganese containing part of the reaction proceeds in the direction as written (ie a reduction half-reaction) and that the iron containing part of the reaction is reversed, ie iron(+2) is oxidised to iron(+3).

$$MnO_4^- + 8H^+ + 5Fe^{2+} \rightleftharpoons Mn^{2+} \; 4H_2O + 5Fe^{3+}$$

In your work always check the balance.

Now at equilibrium

$$E_{(\text{iron half-reaction})} = E_{(\text{manganese half-reaction})}$$

therefore

$$E^{o}_{Fe} - \frac{0.0591}{5} \log \frac{[Fe^{2+}]^5}{[Fe^{3+}]^5} \rightleftharpoons E^{o}_{Mn} - \frac{0.0591}{5} \log \frac{[Mn^{2+}]}{[MnO_4^-][H^+]^8}$$

From the full equation we note that the equilibrium constant is given by;

$$K_{eq} = \frac{[Mn^{2+}][Fe^{3+}]^5}{[MnO_4^-][Fe^{2+}]^5[H^+]^8}$$

Be careful in the rearrangement to ensure that you don't end up with a negative value for the difference of the E^o values, and that in handling the manipulation of the log terms you ensure that K_{eq} is the right way up.

So rearranging we get,

$$E^{o}_{Mn} - E^{o}_{Fe} = \frac{0.0591}{5} \log \frac{[Mn^{2+}][Fe^{3+}]^5}{[MnO_4^-][H^+]^8[Fe^{2+}]^5}$$

ie $\log K_{eq} = \dfrac{0.73 \times 5}{0.0591}$

$$K_{eq} = 5.75 \times 10^{61}$$

****

SAQ 5.2c

> Calculate the equilibrium constant for the reaction of metallic zinc with copper sulphate solution.
>
> $E^{\circ}_{Cu^{2+}/Cu} = + 0.337 \text{ V}$
>
> $E^{\circ}_{Zn^{2+}/Zn} = - 0.763 \text{ V}$

Response

$$K_{eq} = 1.68 \times 10^{37}$$

If you do not agree with this check your working against the previous example; remember that in this case $n = 2$ and that here the signs of E° are different.

****

SAQ 6.1a

> Consider the example above where the unsaturated-oil component of a commercial vegetable oil is determined either by a uv method or by iodometric titration. Assuming several different unsaturated components are present:
>
> (*i*) What would be the value of an 'iodine number' (measure of the unsaturation) obtained volumetrically?
>
> (*ii*) Why would the uv measurement be less useful?
>
> (*iii*) Suggest a method by which all the unsaturated components in the oil might be determined.

Response

(*i*) The iodine number is easily obtained at low cost and is related to the total unsaturation in the sample. This is a useful measurement and, with experience, is good enough for quality control and customer relations purposes. It is hard to relate to percentage composition for complex mixtures.

(*ii*) The different unsaturated components show different absorptions of uv light (E_{max} values) and their absorptions will overlap. The results although extremely precise (often components can be determined to 0.001%), are meaningless unless the nature of the components is known and the mixture is simple.

(*iii*) Gc or hplc separation of components and their determination by uv spectrophotometry would provide greater and more detailed information, so that a full profile of the oil would become available. Considerable capital expenditure on the instrumentation, columns, and method development would be necessary.

This means in practice that the 'iodine value' is still routinely measured, although sophisticated laboratories may opt for the full product-profile.

****

SAQ 6.1b

> 25.00 cm^3 of a solution of Ni(II) and Ca(II) were titrated with 0.1025*M*-EDTA at pH5 and required 22.35 cm^3. The pH was then adjusted to 10 and titration required 14.75 cm^3 EDTA. What is the concentration of Ca(II) and of Ni(II)?
>
> Data: $A_r(Ni) = 58.71$, $A_r(Ca) = 40.08$
>
> at pH5, $Ni^{2+} + EDTA = NiH_2EDTA + 2H^+$
>
> at pH10, $Ca^{2+} + EDTA = CaEDTA^{2-} + 4H^+$

Response

For the Ni(II) solution, number of moles of EDTA

$$= \frac{22.35}{1000} \times 0.1025 = 0.002291 \text{ mole.}$$

Since EDTA reacts 1 : 1 with Ni(II) and A_r Ni = 58.71,

concentration of Ni(II) = 58.71 × 0.002291 = 0.1345 g per 25.00 cm^3.

∴ concentration of Ni(II) is

$$0.1345 \times \frac{1000}{25.00} = 5.380 \text{ g Ni(II) per litre}$$

For the Ca(II) solution number of moles EDTA

$$= \frac{14.75}{1000} \times 0.1025 = 0.001512 \text{ mole.}$$

Ca also reacts 1 : 1 and A_r Ca = 40.08,

∴ concentration of Ca(II) = 0.001512 × 40.08 = 0.06060 g per 25 cm^3, or

$$0.06060 \times \frac{1000}{25} = 2.424 \text{ g Ca(II) per litre.}$$

****

SAQ 6.1c

50.00 cm^3 of a Fe(II)/Fe(III) mixed solution were reduced in a Jones reductor to give Fe(II). Before reduction the Fe(II) required 18.34 cm^3 of 0.02000*M*-$KMnO_4$; after reduction 42.34 cm^3 were needed. ⟶

SAQ 6.1c (cont.)

> **Calculate the concentration of Fe(II) and of Fe(III) in the solution.**
>
> **Data:**
>
> $$Fe^{2+} \rightarrow Fe^{3+} + e$$
>
> $$MnO_4^- + 8H^+ + 5e \rightarrow Mn^{2+} + 4H_2O$$

Response

Fe(II) is 0.0367 *M*, Fe(III) is 0.0480 *M*

18.34 cm^3 of 0.02000*M* $KMnO_4$ contains $\frac{18.34}{1000} \times 0.02$ moles of $KMnO_4$.

As one mole of MnO_4^- reacts with 5 moles Fe(II),

$$\text{moles Fe(II) at start} = \frac{18.34}{1000} \times 0.02000 \times 5$$

$$= 1.834 \times 10^{-3}.$$

This is in 50 cm^3.

$$\therefore \quad \text{molarity of Fe(II) solution} = \frac{1000}{50} \times 1.834 \times 10^{-3}$$

$$= 3.668 \times 10^{-2} M$$

After reduction:

$$\text{Total number of moles Fe(II)} = \frac{42.34 \times 0.02000 \times 5}{1000}$$

$$= 4.234 \times 10^{-3}.$$

$\therefore$ number of moles Fe(III) before reduction

$$= (4.234 - 1.834) \times 10^{-3}$$

$$= 2.400 \times 10^{-3}.$$

This is in 50 cm^3.

$\therefore$ molarity of Fe(III) solution

$$= \frac{1000}{50} \times 2.40 \times 10^{-3}$$

$$= 4.800 \times 10^{-2} M.$$

****

SAQ 6.1d

The chromium in 1.0254 g of an ore was oxidised to $Cr_2O_7^{2-}$ which was treated with 25.00 cm^3 of 0.4000M- Fe(II) solution (an excess). This excess of Fe(II) required 34.85 cm^3 of 0.02642M-$KMnO_4$. What is the percentage of chromium in the ore?

Data: A_r (Cr) = 52.00

$$Fe^{2+} \rightarrow Fe^{3+} + e$$

$$MnO_4^- + 8H^+ + 5e \rightarrow Mn^{2+} + 4H_2O$$

$$Cr_2O_7^{2-} + 14H^+ + 6e \rightarrow 2Cr^{3+} + 7H_2O$$

Response

% chromium in ore = 9.12

The total number of moles Fe(II) $= \dfrac{25.00}{1000} \times 0.4000$

$= 1.0000 \times 10^{-2}$.

The excess of Fe(II) reacted with 34.85 cm^3 of 0.02642M-$KMnO_4$

$$= \frac{34.85 \times 0.02642}{1000} = 9.207 \times 10^{-4} \text{ moles } KMnO_4$$

This is equivalent to $5 \times 9.207 \times 10^{-4} = 4.604 \times 10^{-3}$ moles Fe(II).

$\therefore$ number of moles Fe(II) used =

$$1.0000 \times 10^{-2} - 4.604 \times 10^{-3} = 5.396 \times 10^{-3}.$$

As 2 Cr(III) (or $Cr_2O_7^{2-}$) are equivalent to 6 Fe(II),

$$\text{number of moles Cr(III)} = \frac{5.396 \times 10^{-3}}{3}$$

$$= 1.7987 \times 10^{-3},$$

and as A_r (Cr) = 52.00,

Mass of chromium $= 1.7987 \times 10^{-3} \times 52.00$ g

$= 9.353 \times 10^{-2}$ g.

$$\% \text{ Cr in the ore} = \frac{9.353 \times 10^{-2}}{1.0254} \times 100 = 9.121.$$

****

SAQ 6.1e

0.2734 g of sodium oxalate was dissolved in water, acidified with H_2SO_4, heated to 70 °C and titrated with potassium permanganate solution (42.68 cm^3). Unfortunately too much was added and the excess of $KMnO_4$ was back titrated with 0.1024*N*-oxalic acid, [ethane-1,2-dioic acid, $(COOH)_2$] and needed 1.46 cm^3. What is the concentration normality of the permanganate solution?

Data:

$$5\,C_2O_4^{2-} + 2\,MnO_4^- + 16\,H^+ \rightarrow 2\,Mn^{2+} + 10\,CO_2 + 8\,H_2O$$

ie $C_2O_4^{2-} \rightarrow 2\,CO_2 + 2e$

$M_r\ (Na_2C_2O_4) = 134.00$

$MnO_4^- + 8\,H^+ + 5e = Mn^{2+} + 4\,H_2O$

Response

Number of equivalents $Na_2C_2O_4$ = number of equivalents $KMnO_4$ − number of equivalents $H_2C_2O_4$.

Equivalent weight of $Na_2C_2O_4 = \frac{1}{2}\,M_r = 134.00/2 = 67.00$ g

$\therefore$ number of equivalents of $Na_2C_2O_4 = 0.2734/67.00 = 0.004081$.

Number of equivalents of $KMnO_4$ (normality *N*) used

$$= \frac{42.68}{1000} \times \text{normality of } KMnO_4 = 0.04268 \times N \text{ equivalents.}$$

Number of equivalents oxalic acid $= \dfrac{1.46}{1000} \times 0.1024$

$= 0.0001495$

$\therefore \quad 0.004081 = 0.04268 \times N - 0.0001495$

or $0.004231 = 0.04268 \times N$.

$\therefore \quad$ Normality of $KMnO_4 = \dfrac{0.004231}{0.04268} = 0.09913N$.

****

SAQ 6.1f

0.4671 g of a solid containing sodium hydrogen carbonate was treated with hydrochloric acid (40.72 cm^3). This acid was standardised against 0.1876 g of anhydrous sodium carbonate which needed 37.86 cm^3 of acid.

What is the normality of the acid, and the percentage sodium hydrogen carbonate in the solid?

Data:

equivalent of $NaHCO_3$ = 84.01, of Na_2CO_3 = 52.99

Response

Normality of HC1 = $0.0935N$.

Percentage of sodium hydrogen carbonate = 68.47

0.1876 g of anhydrous Na_2CO_3 contains $\dfrac{0.1876}{52.99} = 3.540 \times 10^{-3}$ equivalents.

$\therefore$ as this reacted with 37.86 cm^3 of acid there must be the same numbers of equivalents of acid, viz 3.540×10^{-3}.

$$\text{ie Normality of acid} = \frac{1000 \times 3.540 \times 10^{-3}}{37.86}$$

$$= 9.350 \times 10^{-2} N.$$

40.72 cm^3 of this acid will contain $\dfrac{40.72}{1000} \times 9.350 \times 10^{-2}$

$= 3.807 \times 10^{-3}$ equivalents.

This is the number of equivalents also in the 0.4671 g of the sodium hydrogen carbonate solid (equivalent 84.01).

$$\therefore \quad \text{mass } NaHCO_3 = 3.807 \times 10^{-2} \times 84.01 \text{ g}$$

$$= 3.198 \times 10^{-1} \text{ g}.$$

$$\% \; NaHCO_3 = \frac{3.198 \times 10^{-1}}{4.671 \times 10^{-1}} \times 100 = 68.47.$$

****

SAQ 6.1g

0.2500 g of pure potassium chloride was mixed with 0.4500 g of impure barium chloride and the mixture was titrated with 0.1000*N*-$AgNO_3$, requiring 72.30 cm^3.

Calculate the percentage purity of the barium chloride.

Data:

Equivalent of KC1 = 74.56, of $BaCl_2$ = 104.12.

Response

Percentage purity of the $BaCl_2$ = 89.71.

72.30 cm^3 of 0.1000*N*-$AgNO_3$ contains

$$\frac{72.30}{1000} \times 0.1000 = 7.230 \times 10^{-3} \text{ equivalents}$$

This will also be the number of equivalents of chloride.

0.2500 g KCl provides $\dfrac{0.2500}{74.56} = 3.353 \times 10^{-3}$ equivalents chloride.

$\therefore$ number of equivalents of chloride from $BaCl_2$

$= (7.230 - 3.353) \times 10^{-3} = 3.877 \times 10^{-3}$.

This equals the number of equivalents of $BaCl_2$.

$\therefore$ mass $BaCl_2$ $= 104.12 \times 3.877 \times 10^{-3}$

$= 4.037 \times 10^{-1}$ g.

$\therefore$ % of $BaCl_2$ in impure $BaCl_2$

$$= \frac{4.037 \times 10^{-1}}{4.500 \times 10^{-1}} \times 100 = 89.71.$$

****

SAQ 6.1h

We wish to make 2*M*-sulphuric acid from commercial conc. sulphuric acid (94.0% w/w and density 1.831 g cm^{-3}) by dilution. How much concentrated acid is needed per litre?

Data: M_r (H_2SO_4) = 98.1.

Response

1.000 g of concentrated acid contains 0.940 g H_2SO_4

which is $\dfrac{0.940}{98.1} = 0.009582$ moles H_2SO_4

This is in a volume of $\dfrac{1.000}{1.831}$ cm^3 = 0.5461 cm^3

$\therefore$ in 1.000 litre the number of moles will be $\dfrac{1000}{0.5461} \times 0.009582$ = 17.55.

Concentration = 17.55 M,

ie a dilution of $\dfrac{17.55}{2}$ is required,

or a volume of acid = $\dfrac{2}{17.55} \times 1000$ cm^3 = 114.0 cm^3 is required.

****

SAQ 6.1i

Calculate the molarity and molality of:

(*a*) 69.0% nitric acid, density 1.409 g cm^{-3},

M_r = 63.01;

(*b*) 85.0% phosphoric acid, density 1.689 g cm^{-3},

M_r = 98.00.

Response

(*a*) 35.3 molal and 15.4 molar.

In 1.00 g of concentrated nitric acid there are 0.69 g of HNO_3 and 0.31 g of H_2O,

ie $\frac{0.69}{63.01}$ moles of HNO_3 per 0.31 g H_2O.

$\therefore$ per 1000 g of water we have

$$= \frac{1000}{0.31} \times \frac{0.69}{63.01} = 35.3 \text{ moles, ie molality} = 35.3.$$

1 g of the acid occupies a volume of $\frac{1}{1.409}$ cm^3

$= 0.7097$ cm^3,

ie $\frac{0.69}{63.01}$ moles of HNO_3 occupy a volume of 0.7097 cm^3.

$\therefore$ in 1 litre of the acid the number of moles of HNO_3

$$= \frac{1000}{0.709} \times \frac{0.69}{63.01} = 15.4, \text{ ie molarity} = 15.4.$$

****

SAQ 6.2a

Suggest methods for comparison with the determination of dilute hydrochloric acid by acid–base titration.

Response

Your answer might include.

silver nitrate titration,
conductimetric titration,
potentiometric titration.

Each might give results differing from the acid–base titration and be superior if the method and equipment were of higher quality.

****

SAQ 6.2b

> **Suppose we made up a solution in a one-litre flask, pipetted 25 cm^3 out of it and titrated it with 25 cm^3 of some other substance from a 50 cm^3 burette.**
>
> **What is the maximum error due to the tolerances given in Fig. 6.2d.**

Response

The graduated flask would give rise to a proportional error of $\frac{25.00}{1000} \times 0.80 = 0.02$ cm^3,

the pipette would have a constant error of 0.06 cm^3, and the burette a proportional error of $\frac{25 \times 0.06}{50} = 0.03$ cm^3,

giving a total maximum error of 0.11 cm^3. The actual error is likely to be less than this.

****

SAQ 6.2c

> **In a typical laboratory the contents (water) of a borosilicate 100 cm^3 pipette were weighed and found to be 99.588 g at 24 °C. What is the error in volume?**

Response

Answer: $-0.045\ cm^3$.

From the table a 100.00 cm^3 pipette measured at 24 °C but quoted at 20 °C, would contain a mass of 99.633 g.

99.588 g is equivalent to a volume of $\dfrac{99.588 \times 100}{99.633}$

$= 99.9548\ cm^3$,

ie a $-0.045\ cm^3$ error.

****

SAQ 7.2a

> **What is the pH of a solution which is**
>
> **(*a*) 4.32×10^{-4} *M* in H^+**
> **(*b*) 5.12×10^{-2} *M* in OH^-?**

Response

(*a*) pH $= \log \dfrac{1}{[H^+]} = \log \dfrac{1}{4.32 \times 10^{-4}}$

$= \log 10^4 - \log 4.32 = 4 - 0.6355$

$= 3.3645$, ie pH $= 3.36$.

(*b*) pOH $= \log \frac{1}{[OH^-]} = \log \frac{1}{5.12 \times 10^{-2}}$

$= \log 10^2 - \log 5.12 = 2 - 0.7093$

$= 1.2907.$

Now $[H^+][OH^-] = 1.00 \times 10^{-14}$ at 25 °C,

$\therefore \quad pH + pOH = 14.0$, ie $pH = 12.7093$

$= 12.71.$

Alternatively,

Since $[H^+][OH^-] = 1.00 \times 10^{-14}$

and $[OH^-] = 5.12 \times 10^{-2}$

$$[H^+] = \frac{1.00 \times 10^{-14}}{5.12 \times 10^{-2}} = 1.9531 \times 10^{-13},$$

whence $pH = \log 10^{13} - \log 1.9531$

$= 13 - 0.2907 = 12.71$ as before.

****

SAQ 7.2b

If pH = 8.32 what is

(*a*) the H^+ concentration,
(*b*) the OH^- concentration?

Response

Again $pH = \log \frac{1}{[H^+]} = 8.32$

$\therefore \quad \log [H^+] = -8.32 = -9 + 0.6800$

$\therefore \quad [H^+] = \text{antilog}(-9 + 0.6800)$

$\qquad = 10^{-9} \times 4.7863$

ie $[H^+] = 4.789 \times 10^{-9}\ M$.

As $[H^+][OH^-] = 1 \times 10^{-14}$

$$[OH^-] = \frac{1.00 \times 10^{14}}{4.79 \times 10^{-9}}$$

$$= 2.09 \times 10^{-6}\ M.$$

****

SAQ 7.2c

Before we leave pH meters and their probes let's revise some of the important practical points. Check your answers with mine and if you get them wrong or are unsure revise and recheck them.

(*a*) Name the four hydrogen-ion indicator electrodes and two main reference electrodes.

(*b*) Which hydrogen-ion indicator electrode is best to use:

(*i*) for strong alkali,
(*ii*) for convenience,
(*iii*) for an effluent from a dye works,
(*iv*) for a fairly uniform nearly neutral factory waste? ⟶

SAQ 7.2c (cont.)

(*c*) What corrections would be applied when using a glass electrode to measure the pH of

(*i*) a sea water sample,
(*ii*) a blood sample,
(*iii*) a viscous polyphosphoric acid sample?

Response

(*a*) The hydrogen, quinhydrone, antimony, and glass electrode. Silver–silver chloride and calomel reference electrodes.

(*b*) (*i*) The hydrogen electrode (unaffected by strong alkali).

(*ii*) The glass electrode (possibly the antimony electrode)

(*iii*) Most likely the glass electrode which is least sensitive to impurities in the effluent.

(*iv*) Most likely the antimony electrode as it is more robust than a glass electrode.

(*c*) (*i*) Corrections for temperature and Na^+ content.

(*ii*) Corrections for ions (blood is salty!), pH is usually quoted at 37 °C for blood samples.

(*iii*) Corrections for activity which may be included in an acid-error correction.

In (*i*) (*ii*) and (*iii*) we would have to allow for *k* the asymmetric junction potential and for the junction potential of the salt bridge.

****

SAQ 7.4a

Calculate the pH at 25 °C of

(*a*) 0.100*M*-NH_4OH,

(*b*) 0.100*M*-NH_4Cl.

[$K_b(NH_4OH) = 1.76 \times 10^{-5}$ at 25 °C]

Response

Making a huge intuitive leap, as the K_b value for NH_4OH is the same as the K_a value for CH_3CO_2H we might predict the values to be 11.12 and 5.12 respectively.

(*a*)

$$NH_4OH \overset{K_b}{\rightleftharpoons} NH_4^+ + OH^-$$

	NH_4OH	NH_4^+	OH^-
initially	c	0	0
At equilibrium	$c - x$	x	x

$$\therefore \quad K_b = \frac{x^2}{(c-x)} \approx \frac{x^2}{c}$$

$$\therefore \quad x = 1.33 \times 10^{-3} M$$

$$\therefore \quad pOH = \log \frac{1}{1.33 \times 10^{-3}} = 2.88.$$

$$\therefore \quad pH = 14 - 2.88 = 11.12.$$

(*b*)

$$NH_4Cl + H_2O \overset{K_a}{\rightleftharpoons} NH_4OH + HCl$$

	NH_4Cl	NH_4OH	HCl
At equilibrium	$c - x$	x	x

$$\therefore \quad K_a = \frac{x^2}{(c-x)} \approx \frac{x^2}{c}$$

$$K_a = \frac{K_w}{K_b} = \frac{1 \times 10^{-14}}{1.76 \times 10^{-5}} = 5.68 \times 10^{-10}$$

$$\therefore \quad x = (5.68 \times 10^{-10} \times 0.1)^{\frac{1}{2}} = 7.55 \times 10^{-6}$$

This is now the $[H^+]$ (or [HCl])

$$\therefore \quad pH = \log \frac{1}{7.55 \times 10^{-6}} = 5.12$$

****

SAQ 7.5a

What is the pH at 25 °C of 0.0200 M-$Na_2C_2O_4$?

(K_{a2} = 5.4 × 10^{-5}
K_w = 1.00 × 10^{-14})

Response

$$C_2O_4^{2-} + H_2O \underset{}{\overset{K_{b1}}{\rightleftharpoons}} HC_2O_4^- + OH^-$$

	$C_2O_4^{2-}$	$HC_2O_4^-$	OH^-
Initially	0.0200	0	0
at equilibrium	$(0.0200 - x)$	x	x

$$\therefore \quad K_{b1} = \frac{x^2}{(0.0200 - x)}$$

Now $K_{b1} = \dfrac{K_w}{K_{a2}} = \dfrac{1.01 \times 10^{-14}}{5.4 \times 10^{-5}}$

$$= 1.87 \times 10^{-10}$$

We can use the approximation.

$$\therefore \quad 1.87 \times 10^{-10} = \frac{x^2}{0.0200}$$

$x^2 = 3.74 \times 10^{-12}$

$x = 1.93 \times 10^{-6}$. This is the $[OH^-]$.

$\therefore$ pOH = 5.71, and as pH + pOH = 14

pH = 8.29.

****

SAQ 7.5b | **What is the pH at 25 °C of 0.100 M-Na_2HPO_4?**

Response

Here the two K values are K_{a2} and K_{a3}.

$K_{a2} = 6.2 \times 10^{-8}$ $K_{a3} = 4.8 \times 10^{-13}$

$K_{a3}\,[HA^-]$ is of the same order as K_w,

but $[HA^-]/K_{a2}$ is $\gg 1$.

$\therefore$ the relationship to use is

$$[H^+] = \left(\frac{K_{a3}[HA^-] + K_w}{[HA^-]/K_{a2}}\right)^{\frac{1}{2}}$$

ie

$$[H^+] = \left(\frac{4.8 \times 10^{-13} \times 0.100 + 1.00 \times 10^{-14}}{0.100/6.2 \times 10^{-8}}\right)^{\frac{1}{2}}$$

$$= \left(\frac{5.80 \times 10^{-14}}{1.61 \times 10^{6}}\right)^{\frac{1}{2}}$$

$$= 1.90 \times 10^{-10}$$

$\therefore$ pH = 9.72.

****

SAQ 7.5c

What is the pH at 25 °C of 0.1000 M-H_3PO_4/0.0200 M-NaH_2PO_4?

($K_{a1} = 7.6 \times 10^{-3}$)

Response

$$H_3PO_4 + H_2O \overset{K_{a1}}{\rightleftharpoons} H_3O^+ + H_2PO_4^-$$

	H_3PO_4	H_3O^+	$H_2PO_4^-$
Initially	0.1000 M	0	0.0200 M
at equilibrium	$0.10 - x$	x	$(0.020 + x)$

Hence $K_{a1} = \dfrac{(x)(0.020 + x)}{(0.10 - x)}$

We cannot approximate as concentration and K_{a1} are close.

$$\therefore \quad 7.6 \times 10^{-3} = \frac{0.02x + x^2}{(0.10 - x)}$$

ie $x^2 + 0.0276x - 0.00076 = 0$

$$\therefore \quad x = \frac{-0.0276 \pm (0.0276^2 + 4 \times 1 \times 0.00076)^{\frac{1}{2}}}{2}$$

$$= \frac{-0.0276 + 6.17 \times 10^{-2}}{2}$$

$= 1.71 \times 10^{-2} M$. Hence pH = 1.77.

****

SAQ 7.5d

What if the pH of the neutralised 0.100 *M*-H_3PO_4 was now 8.0?

$(H^+ = 1.0 \times 10^{-8} \quad K_{a1} = 7.6 \times 10^{-3}$

$K_{a2} = 6.2 \times 10^{-8} \quad K_{a3} = 4.8 \times 10^{-13})$.

Response

$(H^+ = 1.0 \times 10^{-8}$, $K_{a1} = 7.6 \times 10^{-3}$, $K_{a2} = 6.2 \times 10^{-8}$,

$K_{a3} = 4.8 \times 10^{-13})$

Now $[H^+]^3 = 1 \times 10^{-24}$ $K_{a1}[H^+]^2 = 7.6 \times 10^{-19}$

$K_{a1}\ K_{a2}[H^+] = 4.71 \times 10^{-18}$

$K_{a1}\ K_{a2}\ K_{a3} = 2.26 \times 10^{-22}$

and $[H^+]^3 + K_{a1}[H^+]^2 + K_{a1}\ K_{a2}[H^+] + K_{a1}\ K_{a2}\ K_{a3}$
$= 5.48 \times 10^{-18}$

and hence

$$\alpha_0 = \frac{1 \times 10^{-24}}{5.48 \times 10^{-18}} = 1.82 \times 10^{-7}$$

Hence $[H_3PO_4] = 1.82 \times 10^{-8} M$.

$$\alpha_1 = \frac{7.6 \times 10^{-19}}{5.48 \times 10^{-18}} = 1.39 \times 10^{-1}\ [H_2PO_4^-] = 1.39 \times 10^{-2}$$

$$\alpha_2 = \frac{4.72 \times 10^{-18}}{5.48 \times 10^{-18}} = 0.86\ [HPO_4^{2-}] = 8.6 \times 10^{-2}$$

$$\alpha_3 = \frac{2.26 \times 10^{-22}}{5.48 \times 10^{-18}} = 4.12 \times 10^{-5}\ [PO_4^{3-}] = 4.12 \times 10^{-6}$$

Hence $[H_3PO_4]$ = trace, $[H_2PO_4^-]$ = 14%, $[HPO_4^{2-}]$ = 86%,

$[PO_4^{3-}]$ = trace (which ties in with Fig. 7.5c).

****

SAQ 7.6a

Calculate the pH of

(*a*) 0.0100 *M*-CH_3CO_2H,

($K_a = 1.76 \times 10^{-5}$)

(*b*) 0.0100*M*-CH_3CO_2H/0.100*M*.$CH_3CO_2^-Na^+$

Response

$$(a)\ K_a = \frac{[H^+][CH_3CO_2^-]}{CH_3CO_2H} = \frac{x^2}{(c - x)} \simeq \frac{x^2}{c}$$

as in previous calculations where

c = initial [acid],

$x = [H^+], [CH_3CO_2^-]$ at equilibrium.

$\therefore\quad x^2 = 0.01 \times 1.76 \times 10^{-5} = 1.76 \times 10^{-7}$

$x = 4.20 \times 10^{-4}$

ie $[H^+] = 4.20 \times 10^{-4} M$

$\therefore\quad pH = 3.38$

(*b*) Using the same equilibrium

$$K_a = \frac{[H^+][CH_3CO_2^-]}{[CH_3CO_2H]}$$

Here $[H^+] = y + [CH_3CO_2^-] = 0.0100 + y$

$[CH_3CO_2H] = (c - y)$

$$\therefore \quad K_a = \frac{y(0.0100 + y)}{c - y}$$

As y is likely to be small compared to 0.0100 or c we can approximate, ie $K_a = \dfrac{y(0.0100)}{c}$

$$\therefore \quad y = 0.0100 \times 1.76 \times 10^{-3} = 1.76 \times 10^{-5}$$

(Note our guess of the order of magnitude of y was justified)

ie pH = 4.75.

This could have been more quickly found from the Henderson–Hasselbach equation.

$$pH = pK_a + \log \frac{[A^-]}{[HA]} \text{ where } \frac{[A^-]}{[HA]} = \frac{0.0100}{0.0100} = 1.00$$

and $pK_a = 4.75$.

****

SAQ 7.6b

> (*a*) What is the pH of 1.00 *M*-Na_2CO_3/1.00 *M*-$NaHCO_3$ solution?
>
> (*b*) What would be the mass of each in 1.00 dm^3 to give a buffer of pH 9.0?
>
> Data $K_{b1} = 2.1 \times 10^{-4}$ $pK_{b1} = 3.7$

Response

(*a*) Our quickest route to this part (but not the only way) is to use the Henderson Hasselbach equation.

$$pH = pK_a + \log \frac{[A^-]}{[HA]}$$

$[HA] = [HCO_3^-]$ and $[A^-] = [CO_3^{2-}]$

where K_a is for the reaction

$$HCO_3^- + H_2O \overset{K_a}{\rightleftharpoons} H_3O^+ + CO_3^{2-}.$$

$\therefore \quad pH = (14 - 3.7) + \log 1$

$= 10.3 \qquad\qquad pH = 10.3.$

(*b*) We wish to have pH of 9.0 From the Henderson–Hasselbach equation:

$$pH = 10.3 + \log \frac{[A^-]}{[HA]} = 10.3 + \log \frac{[CO_3^{2-}]}{[HCO_3^-]}$$

This means $\log \frac{[CO_3^{2-}]}{[CO_3^{2-}]}$ has to be -1.3

ie $\log \frac{[HCO_3^-]}{[CO_3^{2-}]} = 1.3$

ie ratio $HCO_3 : CO_3 = 10^{1.3} : 1$

$= 19.95 : 1$, ie $20 : 1$.

This means on a molar basis we must have 20 $NaHCO_3$ to 1 Na_2CO_3. We should realise that it is the ratio that matters primarily, so that for the moment any masses fitting this mole ratio will do.

eg $M/10$ M_r $NaHCO_3 = 84$

$\therefore$ $M/10 = 8.4$ g/dm^3

$Na_2CO_3 = 106$

$\therefore$ $M/10 = 10.6$ g/dm^3

We need a ratio of 20 : 1 ie 168 g $NaHCO_3$: 10.6 g Na_2CO_3 per dm^3. Any possible mixture with this ratio will do.

Units of Measurement

For historic reasons a number of different units of measurement have evolved to express quantity of the same thing. In the 1960s, many international scientific bodies recommended the standardisation of names and symbols and the adoption universally of a coherent set of units—the SI units (Système Internationale d'Unités)—based on the definition of five basic units: metre (m); kilogram (kg); second (s); ampere (A); mole (mol); and candela (cd).

The earlier literature references and some of the older text books, naturally use the older units. Even now many practicing scientists have not adopted the SI unit as their working unit. It is therefore necessary to know of the older units and be able to interconvert with SI units.

In this series of texts SI units are used as standard practice. However in areas of activity where their use has not become general practice, eg biologically based laboratories, the earlier defined units are used. This is explained in the study guide to each unit.

Table 1 shows some symbols and abbreviations commonly used in analytical chemistry. Table 2 shows some of the alternative methods for expressing the values of physical quantities and the relationship to the value in SI units.

More details and definition of other units may be found in the *Manual of Symbols and Terminology for Physicochemical Quantities and Units*, Whiffen, 1979, Pergamon Press.

Table 1 *Symbols and Abbreviations Commonly used in Analytical Chemistry*

Symbol	Meaning
Å	Angstrom
$A_r(X)$	relative atomic mass of X
A	ampere
E or U	energy
G	Gibbs free energy (function)
H	enthalpy
J	joule
K	kelvin (273.15 + t °C)
K	equilibrium constant (with subscripts p, c, therm etc.)
K_a, K_b	acid and base ionisation constants
$M_r(X)$	relative molecular mass of X
N	newton (SI unit of force)
P	total pressure
s	standard deviation
T	temperature/K
V	volume
V	volt ($J\ A^{-1}\ s^{-1}$)
a, a(A)	activity, activity of A
c	concentration/ mol dm^{-3}
e	electron
g	gramme
i	current
s	second
t	temperature / °C
bp	boiling point
fp	freezing point
mp	melting point
$\approx$	approximately equal to
$<$	less than
$>$	greater than
e, exp(x)	exponential of x
ln x	natural logarithm of x; $\ln x = 2.303 \log x$
log x	common logarithm of x to base 10

Table 2 *Alternative Methods of Expressing Various Physical Quantities*

1. **Mass (SI unit : kg)**

$$g = 10^{-3}\ kg$$
$$mg = 10^{-3}\ g = 10^{-6}\ kg$$
$$\mu g = 10^{-6}\ g = 10^{-9}\ kg$$

2. **Length (SI unit : m)**

$$cm = 10^{-2}\ m$$
$$Å = 10^{-10}\ m$$
$$nm = 10^{-9}\ m = 10Å$$
$$pm = 10^{-12}\ m = 10^{-2}\ Å$$

3. **Volume (SI unit : m^3)**

$$l = dm^3 = 10^{-3}\ m^3$$
$$ml = cm^3 = 10^{-6}\ m^3$$
$$\mu l = 10^{-3}\ cm^3$$

4. **Concentration (SI units : mol m^{-3})**

$$M = mol\ l^{-1} = mol\ dm^{-3} = 10^3\ mol\ m^{-3}$$
$$mg\ l^{-1} = \mu g\ cm^{-3} = ppm = 10^{-3}\ g\ dm^{-3}$$
$$\mu g\ g^{-1} = ppm = 10^{-6}\ g\ g^{-1}$$
$$ng\ cm^{-3} = 10^{-6}\ g\ dm^{-3}$$
$$ng\ dm^{-3} = pg\ cm^{-3}$$
$$pg\ g^{-1} = ppb = 10^{-12}\ g\ g^{-1}$$
$$mg\% = 10^{-2}\ g\ dm^{-3}$$
$$\mu g\% = 10^{-5}\ g\ dm^{-3}$$

5. **Pressure (SI unit : N m^{-2} = kg m^{-1} s^{-2})**

$$Pa = Nm^{-2}$$
$$atmos = 101\ 325\ N\ m^{-2}$$
$$bar = 10^5\ N\ m^{-2}$$
$$torr = mmHg = 133.322\ N\ m^{-2}$$

6. **Energy (SI unit : J = kg m^2 s^{-2})**

$$cal = 4.184\ J$$
$$erg = 10^{-7}\ J$$
$$eV = 1.602 \times 10^{-19}\ J$$

Table 3 *Prefixes for SI Units*

Fraction	Prefix	Symbol
10^{-1}	deci	d
10^{-2}	centi	c
10^{-3}	milli	m
10^{-6}	micro	μ
10^{-9}	nano	n
10^{-12}	pico	p
10^{-15}	femto	f
10^{-18}	atto	a

Multiple	Prefix	Symbol
10	deka	da
10^{2}	hecto	h
10^{3}	kilo	k
10^{6}	mega	M
10^{9}	giga	G
10^{12}	tera	T
10^{15}	peta	P
10^{18}	exa	E

Table 4 *Recommended Values of Physical Constants*

Physical constant	Symbol	Value
acceleration due to gravity	g	9.81 m s^{-2}
Avogadro constant	N_A	6.022 05 × 10^{23} mol^{-1}
Boltzmann constant	k	1.380 66 × 10^{-23} J K^{-1}
charge to mass ratio	e/m	1.758 796 × 10^{11} C kg^{-1}
electronic charge	e	1.602 19 × 10^{-19} C
Faraday constant	F	9.648 46 × 10^{4} C mol^{-1}
gas constant	R	8.314 J K^{-1} mol^{-1}
'ice-point' temperature	T_{ice}	273.150 K exactly
molar volume of ideal gas (stp)	V_m	2.241 38 × 10^{-2} m^3 mol^{-1}
permittivity of a vacuum	ϵ_0	8.854 188 × 10^{-12} kg^{-1} m^{-3} s^4 A^2 (F m^{-1})
Planck constant	h	6.626 2 × 10^{-34} J s
standard atmosphere pressure	p	101 325 N m^{-2} exactly
atomic mass unit	m_u	1.660 566 × 10^{-27} kg
speed of light in a vacuum	c	2.997 925 × 10^{8} m s^{-1}